AS/A2 Chemistry

Contents

Introduction

■ ■ ■

Practical Tasks

Qualitative tasks

Exemplar qualitative tasks

Quantitative tasks

Exemplar quantitative tasks

Evaluative tasks

Exemplar evaluative tasks

OCR(A) AS A2

UNITS
F323
F326

Chemistry

Practical Skills in Chemistry

John Older

Philip Allan Updates, an imprint of Hodder Education, an Hachette UK company, Market Place, Deddington, Oxfordshire OX15 0SE

Orders

Bookpoint Ltd, 130 Milton Park, Abingdon, Oxfordshire OX14 4SB
tel: 01235 827827
fax: 01235 400401
e-mail: education@bookpoint.co.uk

Lines are open 9.00 a.m.–5.00 p.m., Monday to Saturday, with a 24-hour message answering service. You can also order through the Philip Allan Updates website: www.philipallan.co.uk

ISBN 978-1-4441-0844-6

First printed 2010
Impression number 5 4 3
Year 2014 2013 2012 2011

This guide has been written specifically to support students preparing for the OCR(A) AS/A2 Chemistry Units F323 & F326 Practical Skills assessments. The content has been neither approved nor endorsed by OCR and remains the sole responsibility of the author.

Typeset by Tech-Set Ltd, Gateshead, Tyne & Wear
Printed in India

Hachette UK's policy is to use papers that are natural, renewable and recyclable products and made from wood grown in sustainable forests. The logging and manufacturing processes are expected to conform to the environmental regulations of the country of origin.

P01613

Introduction

About this guide

This guide is written for students studying the OCR(A) AS/A2 Chemistry course. It provides advice on the AS and A2 practical skills in chemistry units F323 and F326. The guide should be used in conjunction with the instructions you get from your teacher.

The guide is structured as follows:

- The introduction gives general advice on how to use this guide and matters to do with practical work.
- The second section gives advice on the qualitative tasks with some example tasks to enable you to test your understanding.
- The third section gives advice about quantitative tasks, with examples.
- The final section gives advice about the evaluative tasks, with examples.

The assessments

At both AS and A2, you are assessed on three different types of task. The **qualitative task** is marked out of 10 and involves doing an experiment, observing what happens, recording the results and making simple inferences from what has been observed. The **quantitative task** is marked out of 15 and involves carrying out an experiment, making careful measurements, recording the results and using them to come to a conclusion. The **evaluative task**, which carries 15 marks, requires you to interpret the results of an experiment and consider weaknesses in the practical procedure used and the measurements made. Methods that could be used to improve the experiment and the effect of any change to the procedure might also be tested. At AS and particularly at A2, an understanding of the theory on which the practical procedures are based is expected. Three different experiments are available for each type of task and, if more than one experiment is attempted in each category, the best mark scored is submitted for the examination.

How to use this guide

This book contains a lot of detailed information and it is not intended that you should read it from cover to cover. However, when doing an experiment in class you may find it helpful to refer to the section of the book in which some of the issues relevant to that type of practical are considered. For example, if you are doing a qualitative exercise, you should check that section of the guide and note the common mistakes that are made. The section on quantitative tasks (pp. 28–38) contains a useful reminder of the correct way to handle pipettes, burettes and volumetric flasks. Only practice will give you the competence level you need but if the apparatus is not used correctly, you will not be able to obtain consistent results. Marks are awarded for recording results correctly and you can use this guide to confirm that you are doing this. The correct use of significant figures is very important, and there is a section that explains good practice (pp. 35–37).

Although this guide covers most of the common errors and misunderstandings, you should not assume that it is comprehensive. The wide variety of possible experiments means that it is not possible to cover every variation in the questions that could be asked. The purpose of the tasks is to make you think carefully about what you do. You will be given plenty of time to complete a task. It is a mistake to rush to describe an observation or put down an answer when a moment's thought would allow you to avoid an unnecessary error. This applies particularly to the evaluative tasks, which, since they may refer to experiments that are less familiar, are the hardest to prepare for and to answer. Don't be frightened by them — they always contain some straightforward questions, which, if you don't panic, you should be able to answer.

Some topics not included

There are some things that you will not find in this book. The background theory of the topics that are assessed can be found in any good textbook. The quantitative task, and often the evaluative task, may contain a calculation and part of your preparation *must* be to make sure you know how to handle the steps involved. In particular, it is worth emphasising that writing equations is likely to appear in all the tasks and you *must* make sure that you know how to write formulae and balance equations without making mistakes.

In the tasks, you will be provided with essential safety information. However, you should also have learned the correct way to handle apparatus and chemicals safely and it is possible that you might be asked to provide some comment that indicates you are familiar with normal precautions.

The exemplar tasks

Three sections of this guide contain examples of the kinds of task you may expect to see in the assessments. These are not past examination questions, as OCR does not release them. However, they are designed to illustrate as closely as possible questions that might be asked. Where an assessment requires you to carry out an experiment, the task will contain full details of the practical procedure to be used. Complete details are not included in the exemplars provided in this book, as they are not always necessary in order to understand the questions that follow. Nevertheless, there is sufficient information to allow you to appreciate what would be involved.

Not all possible topics are covered by the exemplars but they do illustrate mistakes that are commonly made and which apply to experiments of a similar type.

Where a task relates specifically to A2 work, this is indicated. However, A2 students should note that many of the issues covered in the AS exemplars are still relevant to their work.

Tackling a task

An important point to emphasise is that you should always read the instructions carefully and then follow them *exactly*. When giving the details of an experiment, great care has been taken by OCR to make sure that everything will work as intended. Do not be tempted to change the procedure in any way — even if you have carried out a similar experiment in class using a slightly different method.

Never give up on a task even if you feel it is not going well. It is possible to gain credit for an answer that is only partially correct. In calculations, if you make a mistake in an early part, it does not mean that you will lose all the marks for the subsequent steps. The evaluative tasks do contain some demanding questions and you may be unsure of the correct response. However, you should always suggest a possible answer and then carry on — often, easier questions follow.

Practical
Tasks

This section describes and explains the three types of tasks you will be assessed on.

- The **qualitative task** requires you to do an experiment, observe what happens, record the results and make simple inferences from your observations.
- The **quantitative task** requires you to carry out an experiment, make careful measurements, record the results and use them to come to a conclusion.
- The **evaluative task** requires you to interpret the results of an experiment and consider weaknesses in the practical procedure used and the measurements made. Methods that could be used to improve the experiment and the effect of any change in the procedure might also be tested.

Qualitative tasks

Students often imagine that the qualitative task will be the most straightforward of the three assessed tasks. It may seem that little skill is required, apart from the ability to follow instructions, write down what is seen and come to simple conclusions. In practice, it is all too easy to carry out an experiment carelessly and fail to notice changes that more careful inspection would have made obvious. Marks are also often lost by the inability to record these observations correctly. To avoid these difficulties careful preparation and attention to detail are necessary.

What knowledge is required?

Knowledge of a small number of practical tests and, in some cases, what can be deduced from them is essential. These are listed below. You are not expected to have a detailed knowledge of the theory on which the tests are based and any deductions that might be required will be straightforward.

AS and A2

The reaction of dilute acids with metals, oxides, hydroxides and carbonates

In each case, you should be prepared to write a balanced equation for the reaction that takes place.

The more reactive metals such as magnesium, iron and zinc react with dilute acids to produce a salt. Hydrogen is given off, so you will see bubbling as the grey metal reacts. A solution of the salt is obtained. For example:

$$Mg(s) + 2HCl(aq) \rightarrow MgCl_2(aq) + H_2(g)$$

It is possible to identify the hydrogen because it gives a 'pop' when a lighted splint is held in the gas.

Most oxides and hydroxides are insoluble in water (except those of group 1 and to a lesser extent calcium hydroxide and barium hydroxide) but they do react with acids to form a salt. What is seen is simply the solid dissolving to form a solution. No gas is given off so there is no bubbling. For example:

$$CuO(s) + 2HCl(aq) \rightarrow CuCl_2(aq) + H_2O(l)$$

$$Mg(OH)_2(s) + H_2SO_4(aq) \rightarrow MgSO_4(aq) + 2H_2O(l)$$

Most carbonates are insoluble in water. When acids are added they bubble vigorously and a salt is formed. The bubbles are carbon dioxide. For example:

$$CaCO_3(s) + 2HNO_3(aq) \rightarrow Ca(NO_3)_2(aq) + CO_2(g) + H_2O(l)$$

Although what you see is similar to the reaction of metals and acids, carbonates are usually easily distinguished by their appearance.

Carbon dioxide can be identified if the gas is bubbled into limewater because a white precipitate is formed. Although the word 'milky' is often used to describe this precipitate, it is best avoided if you are asked to record this as an observation.

In all the cases given above, the acid is neutralised during the course of the reaction. This can be observed if an acid–base indicator is used.

The effect of heat on carbonates (thermal decomposition)

The carbonates of group 1 metals do not decompose on heating, but all the others do. In all cases, an oxide is formed and carbon dioxide is given off. For example:

$$MgCO_3(s) \rightarrow MgO(s) + CO_2(g)$$

Often the temperature has to be quite high to start the decomposition and sometimes, because there is no colour change, it is not obvious that anything has occurred. For example, all group 2 carbonates and oxides are white. However, copper(II)carbonate is green and copper(II) oxide is black, so this would be observed easily. Zinc carbonate is unusual in that it is white when cold but pale yellow when hot.

The reaction of aqueous silver nitrate with solutions of chlorides, bromides and iodides

In all three cases, a precipitate of the silver halide is obtained. However, the precipitates have different colours: silver chloride is white, silver bromide is cream and silver iodide is yellow. Be careful to let the precipitates settle before deciding on the colour, particularly if the solution from which they are obtained is coloured. The colour should allow you to tell the precipitates apart but, since the colours are fairly similar, the addition of aqueous ammonia is sometimes used:

- Silver chloride dissolves in dilute aqueous ammonia.
- Silver bromide dissolves in concentrated aqueous ammonia.
- Silver iodide does not dissolve.

You might be asked to write full or ionic equations for these reactions. For example:

$$AgNO_3(aq) + NaCl(aq) \rightarrow AgCl(s) + NaNO_3(aq)$$
$$Ag^+(aq) + Cl^-(aq) \rightarrow AgCl(s)$$

The reactions of halogens and halide ions

Chlorine reacts with bromides and bromine is formed. This appears as either a yellow colouration if the concentration of bromine is low or brown if more bromine is formed. Again, either a full equation or an ionic equation might be expected — for example:

$$Cl_2(aq) + 2KBr(aq) \rightarrow Br_2(aq) + 2KCl(aq)$$
$$Cl_2(aq) + 2Br^-(aq) \rightarrow Br_2(aq) + 2Cl^-(aq)$$

Chlorine reacts with iodides to form iodine. The colour observed can be confusing because the iodine formed may appear to be brown, grey or purple depending on the concentrations of the solutions.

$$Cl_2(aq) + 2I^-(aq) \rightarrow I_2(aq) + 2Cl^-(aq)$$

Bromine also reacts with iodides to form iodine. This is a reaction that can be easily missed because the brown bromine may produce a brown solution of iodine.

Two reactions of organic compounds

- If drops of bromine water are added to an alkene, the bromine loses its colour. This is a useful test to show that a compound contains a C=C double bond (i.e. it is unsaturated). The reaction between ethene and bromine is an example:

$$CH_2CH_2 + Br_2 \rightarrow CH_2BrCH_2Br$$

- When a primary or secondary alcohol is warmed with an acidified solution of potassium dichromate the colour changes from orange to green. A primary alcohol is oxidised first to an aldehyde and then to a carboxylic acid. A secondary alcohol is oxidised to a ketone. Tertiary alcohols are not oxidised, so no colour change is observed. If you are asked to write an equation you should use [O] to represent the oxidising agent. The oxidation of ethanol (primary alcohol) to ethanoic acid is an example:

$$CH_3CH_2OH + 2[O] \rightarrow CH_3COOH + H_2O$$

A2 only

Further organic reactions

- Phenols, like alkenes, react with bromine water and decolorise it. Unlike alkenes, when sufficient bromine is added a white precipitate is formed. This can be used to distinguish phenols from other types of alcohol.
- Compounds that contain a carbonyl functional group can be distinguished from other compounds because they react with a solution of 2,4-dinitrophenylhydrazine (2,4-DNPH) to produce a yellow or orange precipitate. For this test to work reliably it is essential to add an *excess* amount of the 2,4-DNPH compared with the carbonyl compound. This is because the precipitate that forms is soluble in carbonyl compounds and may dissolve if too much of the latter is used.
- Aldehydes can be distinguished from ketones using Tollens' reagent. When an aldehyde is warmed gently in a water bath with Tollens' reagent, a precipitate of silver is produced. If all the glassware used is clean this appears as a silver 'mirror' although more often the precipitate is observed as a grey powder. This is a redox reaction and ketones do not react because they cannot be oxidised.

Recognising the type of reaction that is taking place

Precipitation

Precipitation occurs when two solutions are mixed and, as a result, an insoluble substance is formed. The first observation is that the contents of the test tube become 'cloudy'. Then, after some time, a solid settles out and the solution above it becomes clear. Transition metal ions form hydroxide precipitates with aqueous sodium hydroxide and aqueous ammonia. An example is the formation of a dark green precipitate of iron(II) hydroxide:

$$FeSO_4(aq) + 2NaOH(aq) \rightarrow Fe(OH)_2(s) + Na_2SO_4(aq)$$

$$Fe^{2+}(aq) + 2OH^-(aq) \rightarrow Fe(OH)_2(s)$$

There are many other cases where precipitation occurs as a wide range of substances are insoluble in water.

Redox reactions

Redox reactions can often be recognised. You should know the following three examples. In each case, the colour change that occurs identifies the reaction as being redox.

- When aqueous potassium iodide is added to another solution and a brown (or grey or purple) colour is observed, this indicates that the iodide has been oxidised to iodine by the other solution.
- When an acidified solution of potassium dichromate changes from orange to green on warming with another substance, this indicates that the potassium dichromate has been reduced and the other substance has been oxidised.
- When an acidified solution of potassium manganate(VII) changes from purple to colourless or very pale pink on warming with another substance, this indicates that the potassium manganate(VII) has been reduced and the other substance has been oxidised.

Ligand substitution reactions

Ligand substitution reactions cannot always be identified easily. A sign that this type of reaction has occurred is that there is a change in colour when two solutions are mixed and there is no other indication that a reaction has taken place. A ligand must have an atom with at least one spare pair of electrons to create the coordinate bond. Therefore, one of the solutions is likely to contain either a molecule or ion with an oxygen, or nitrogen or halogen atom. Ammonia is an example of a common ligand but there are many others. A reaction accompanied by a colour change could, of course, be redox so ligand exchange should not be the conclusion if aqueous solutions of potassium iodide, potassium dichromate or potassium manganate(VII) have been used.

An example of ligand exchange is the addition of concentrated hydrochloric acid to a blue solution containing $Cu(H_2O)_6^{2+}$ ions. The colour steadily changes to green as the H_2O ligands are replaced by Cl^- ligands. (Although it is worth noting that the green colour is actually the result of a mixture of blue $Cu(H_2O)_6^{2+}$ ions and yellow $CuCl_4^{2-}$ ions.)

Some aqueous transition metal ions (and a few others) form a precipitate when aqueous sodium hydroxide or aqueous ammonia is added, but this dissolves as more reagent is added. This is because a complex ion is formed. For example, with aqueous ammonia, copper ions form a pale-blue precipitate of copper(II) hydroxide with the $OH^-(aq)$ present.

$$NH_3(aq) + H_2O(l) \rightleftharpoons NH_4^+(aq) + OH^-(aq)$$

$$Cu^{2+}(aq) + 2OH^-(aq) \rightarrow Cu(OH)_2(s)$$

The precipitate dissolves when more aqueous ammonia is added and a deep-blue solution containing the ion $[Cu(NH_3)_4(H_2O)_2]^{2+}$ is formed.

What skills are required?

During your A-level course you will master various practical skills The qualitative assessment may test that you have achieved a suitable level of competence in manipulating apparatus by providing instructions for a procedure that must be carried out successfully. This could involve making a solution or preparing a solid, either as a powder or as crystals. There are a number of skills that might be required.

Measuring amounts of solids and volumes of solutions

In the tasks, quantities may be expressed in approximate amounts, but it is important not to take this to mean that it doesn't really matter. Sometimes it may be difficult to make the observation that the examiner is expecting if the quantities used are too different from those specified.

The amount of solid may be expressed in terms of a certain number of 'spatulas full'. This is, of course, rather vague as spatulas differ in size but, in general, use less solid rather than more. It is a common mistake to think that larger quantities will make it easier to see what happens.

You may be asked to add volumes of solutions as a measured depth in a test tube or boiling tube. Before you tackle an assessment, check that you can roughly recognise what a depth of 1 cm or 2 cm looks like.

If you are asked to add drops of solution it is easy to add the drops too quickly. To avoid this, it is usually better to hold the dropping pipette vertically and to press the bulb very gently.

Heating a substance

Heating a substance should never be carried out using the yellow flame of a Bunsen burner. The luminous blue flame should be used at first, so that the substance is heated gently. Use the blue-cone flame once the container has been warmed.

Filtering

First, make sure that you know how to fold filter paper correctly. Even then, if the solid that is being removed has very fine particles, the filter paper may become clogged. If this occurs, filtration will be very slow. If it is the filtrate that you need, let the solid settle in its container before pouring the solution through the filter paper. If it is important to obtain all the solid, it will be necessary to carefully wash the last bits of the solid into the filter paper using a jet from a wash bottle containing distilled water.

At A2, you might be asked to filter under reduced pressure and you should check that you are familiar with the particular apparatus used in your laboratory. This is a likely procedure if you are preparing a solid organic product.

Forming crystals

Obtaining good crystals from a solution is surprisingly difficult. It is usually done by heating a solution until it becomes so concentrated that the crystals will start to form as soon as the solution cools. However, it is difficult to tell at what point to stop heating. It is important that the heating is reasonably gentle, so do not use the blue-cone flame. Stop heating immediately the solution begins to spit or there is any sign of solid forming just above the surface of the solution.

At A2, you may be asked to recrystallise a contaminated solid to obtain pure crystals. This is a normal procedure when making a solid organic product. To do this, the impure substance must be dissolved in the *minimum* amount of a warmed solvent (usually water) and then left for the crystals of the product to precipitate out as the solution cools. The important word here is 'minimum'. It is worth struggling to get the impure solid to dissolve in the least possible amount of solvent by stirring it continuously. If too much solvent is used the pure crystals may be extremely reluctant to reform. The purpose of recrystallisation is:

- first, to extract the desired product from any impurities that do not dissolve in the solvent

and

- second, to remove any impurities that are very soluble in the solvent — these will stay dissolved while the desired product recrystallises

Although it is unlikely that you will be asked to do this, the purity of a solid can be confirmed by taking its melting point. A pure solid melts at a fixed temperature. A contaminated product begins to melt at a temperature below the true melting point and the melting then takes place over a wide range of temperatures.

Making observations

Before you start any part of an experiment you should consider what is likely to occur. This may sound impossible but, in practice, only a limited number of changes can occur.

Heating a solid

Any one of three things may be observed:

- The solid may decompose and give off water vapour or a gas.
- The solid may change colour.
- The solid may change state. It may melt to a liquid or, in a limited number of cases, it may sublime (turn to a vapour without passing through a liquid phase).

Always take a close look at the substance before you start heating it. If it is in the form of crystals it is likely (though not a certainty) that water vapour will be given off when it is heated. This will be obvious as the vapour will be visible and droplets of water are likely to form at the top of the tube.

Gases may be given off. A few gases are coloured, but as this is relatively unusual the gas may not be easy to identify. Some, like oxygen (which relights a glowing splint) or hydrogen (which gives a 'pop' with a lighted splint), do have a specific test, but you should expect the instructions to advise you to do these tests if they are required. Be careful to consider obvious things, such as the test for oxygen is unlikely to work if water vapour is being produced at the same time. Testing with an indicator paper such as litmus or pH paper will give an indication as to whether a gas is acidic or alkaline but will not allow you to identify it positively. Don't forget to smell the gas, but do this cautiously as some gases can be choking.

A substance may change colour when heated. This should be clearly visible. Remember to record both the initial and the final colour of the solid.

Some solids melt when they are heated by a Bunsen flame. If this occurs readily then it is likely that the substance is organic because covalent compounds melt easily. Ammonium compounds also melt on heating but they decompose readily and can be identified by the characteristic smell of ammonia.

A few compounds that contain a metal (ionic compounds) can be melted, but this requires continuous heating with a very hot flame. Some crystals may appear to have melted but what is happening is that the substance is dissolving into the water released from the crystals as they are warmed. If you are uncertain whether this has occurred just record the fact that a liquid is formed.

Sometimes a change can occur as the substance cools. An instruction would tell you to keep the substance for a while after heating and observe what happens. Make sure that you leave it for as long as possible as hot tubes cool only slowly.

Mixing two solutions

Sometimes nothing is observed when two solutions are mixed but this does not mean that nothing can be concluded. A test of this kind might serve to eliminate various options when a substance is being identified. Otherwise, there are three possible observations:

- A precipitate may be formed.
- A colour change may occur.
- A gas may be given off.

A precipitate is a solid formed as a result of the mixing of solutions. It always occurs if the mixture contains the components of an insoluble substance. For example, silver chloride is insoluble in water so it forms as a precipitate when aqueous silver nitrate is added to aqueous magnesium chloride — the silver nitrate provides the silver ion and the magnesium chloride provides the chloride ion. It is not necessary to learn a list of which substances are insoluble in water, but you should know the tests for the halides (chloride, bromide and iodide) (see p. 12). Some precipitates are white; others are coloured.

Sometimes a colour change may be observed on mixing two solutions, although no precipitation occurs. At A2, this usually indicates either a redox reaction or a ligand substitution.

If a gas is released when solutions are mixed, it will be obvious because rapid bubbling will be observed. In fact, care often needs to be taken as the gas produced causes the solutions to rise up the tube. It might be possible to test for the gas. For example, carbon dioxide produces a white precipitate of calcium carbonate when passed into limewater.

Adding a solution or a liquid to a solid

There are two possibilities:

- The substance may simply dissolve with nothing further occurring.
- A gas may be given off and the substance may dissolve.

In both cases there could also be a colour change.

Where the substance just dissolves and no further reaction is apparent many students will describe this as 'no reaction'. This is not strictly accurate and, although an examiner might accept it, it is better to describe exactly what is seen — for example, the white solid dissolves to give a colourless solution.

When a solid is added to a liquid it is usual for a few bubbles to be seen as a small amount of air is displaced. There may also be a few bubbles that sit on the surface of the solid. There is no need to record this and it is not evidence of a reaction. A reaction produces a constant stream of bubbles. This will be obvious. However, students often miss the fact that, almost always, the solid dissolves at the same time. You should not forget to include this when you describe the reaction.

Recording observations

As with the examination papers, the space provided will be more than sufficient to record all the observations that are expected. You are not required to write down great detail and it is the key points that gain the marks. You do not have to write in sentences; you should record what you saw simply and briefly. Do not muddle a deduction with an observation. For example, you might decide that when you added a dilute acid to a substance that carbon dioxide was given off — that is a deduction. You will have seen bubbles being produced and the solid dissolving. So you should put that as the observation and carbon dioxide as the deduction (if this has been asked for).

Using the correct word to describe what you have seen is important. So if you mix two solutions and the tube contents become 'cloudy' or 'murky', remember that this means that a precipitate has been formed. This is what you should write down. It is well known that people see colours in different ways, so the examiner will allow some variation. However, you should settle for a single colour rather than attempting to cover all possibilities by writing descriptions such as 'yellowy orange-brown'.

Common mistakes

There are a number of mistakes that are frequently made by students. These include a failure to follow the instructions provided, poor use of apparatus and careless observation or inaccurate recording.

Failure to follow instructions carefully

The experiments you are asked to do have been tested to make sure that the procedure works reliably and that you will be able to see what is expected. The success of many experiments depends on both the conditions used and the quantities of reagents involved. You should take care with the following points:

- Use the volumes specified. These may be given as a particular depth in a test tube or boiling tube.
- When adding one reagent to another *always* do so slowly, even if the instructions do not specify this. A reaction may occur rapidly and be missed or sometimes vigorous bubbling could cause a tube to overflow.
- When asked to add drops of solution do not interpret this as 'squirts'. Learn to use a teat pipette correctly so that this doesn't happen. There are some occasions when a small volume of an added solution causes a particular reaction to occur while a larger volume produces a different result. A2 students should take particular care if asked to add either aqueous sodium hydroxide or aqueous ammonia.
- Take care when heating solids. Do *not* use a very hot flame to start with. There is the obvious risk of breaking glass, but it is also easy to miss a change in the solid that occurs straightaway.
- Be patient when filtering a solid. First, fold the filter paper carefully. Then add the solution containing the precipitate so that it does not overflow the sides of the filter paper and drip down the inside of the funnel. Finally, do not stir or prod the contents of the filter paper with a glass rod or a spatula in the hope of speeding the filtering process. This can lead to the paper tearing and the whole of its contents, including the precipitate, passing through into the container below.

Careless observing and recording

There is a wide range of errors that can be made in observing and recording. Since the majority of marks are awarded for these aspects, particular attention should be paid to avoiding the following mistakes:

- Most observations are easy to make, but be careful if a precipitate is formed from solutions that are coloured. It can be difficult to be sure of the true colour of a solid if it is suspended in a coloured solution. For example, a precipitate of copper(II) hydroxide is very pale blue but, when it is within a solution of copper(II) sulfate, it appears to be quite dark. So, you should let the precipitate settle to the bottom of the tube before deciding on its colour. This may take some time, but you can leave it to stand while you carry on with another experiment.

- Record a colour change rather than just the colour produced. When it reacts, potassium dichromate may change from orange to green. So don't just write 'goes green'.
- Don't use the word clear when you mean colourless. All true solutions are clear, although only some are coloured. So if you want to record that the solution has no colour, write 'colourless'.
- If a solid dissolves when water is added, then write this down rather than writing 'no reaction'.
- Solids tend to produce a small number of bubbles when added to a liquid. Don't record this as bubbling or effervescence. Unless the bubbling is reasonably continuous it is unlikely that a reaction is taking place.

Exemplar qualitative tasks

The tasks below are typical of those you might expect in the qualitative assessment, although in the assessment you will be given more detailed instructions as to exactly how to carry out the tests. The observations can only be practised in class but the following illustrations of students' work focus on the errors to be avoided when recording results.

AS task 1: reactions of unknown solutions

The candidates were given four unknown solutions, A, B, C and D. The solutions were:

A   potassium iodide

B   copper sulfate

C   copper chloride

D   potassium bromide

Candidates were asked to record their observations when the following tests were carried out:

(1) Add aqueous silver nitrate to each of A, B, C and D, followed by an excess of dilute aqueous ammonia. (2 + 1 + 2 + 1 marks)

(2) Add an aqueous solution of chlorine to each of A, B, C and D. (4 marks)

Total: 10 marks

■ ■ ■

Candidate's response

(1)

	Addition of $AgNO_3$(aq)	After aqueous ammonia is added
Solution A	Goes yellow	Stays the same
Solution B	Nothing	Nothing
Solution C	Pale blue solid formed	Solid disappears
Solution D	Goes milky	Milky appearance spreads out

Addition of $AgNO_3$(aq) to Solution A produces a yellow precipitate. The correct words must be used. Writing 'goes yellow' is not sufficient for the first mark. There is no reaction when ammonia is added, so the comment 'stays the same' gains the mark.

Solution B does not react with aqueous silver nitrate. It would be better to write 'no observable reaction' but 'nothing' would gain the mark.

Solution C produces a white precipitate of silver chloride. The candidate has looked at the precipitate through the blue solution of the copper salt and has, therefore, not observed the colour correctly. This is an easy mistake to make but, nevertheless, the candidate loses a mark. The precipitate does dissolve when ammonia is added, so a mark is awarded for 'solid disappears'.

Solution D produces a cream precipitate. However, the student has decided to describe this as 'milky'. This is not allowed as a description of a precipitate, so the candidate fails to score.

(2)

	Addition of Cl_2(aq)
Solution A	Bubbles and brown solution formed
Solution B	Nothing
Solution C	Nothing
Solution D	Brown solution formed

A brown solution is a correct observation for A. (Iodine in solution often appears to be brown if potassium iodide has been used.) But the candidate has imagined that bubbles have been produced when none is formed. The mark would not be awarded. It is a common error to think bubbles are being produced when chemicals are mixed. For it to be an observation worth recording, a sustained flow of bubbles must be observed.

The observations for B, C and D are correct, although once again it would be better to write 'no observable reaction', rather than 'nothing'.

Overall, the candidate scores 6 marks out of 10. This is not disastrous, but it could easily have been better if a little more care had been taken.

AS task 2: reactions of a mixture of two unknown powders

Each candidate was provided with a solid mixture of magnesium iodide and magnesium carbonate. Candidates were told that the mixture contained two compounds of the same metal, but they were not told what the mixture was. They were asked to carry out the following sequence of reactions. In each case, they were asked to record their observations and, where possible, to suggest likely conclusions from their observations.

(1) Place a 3 cm depth of hydrochloric acid in a test tube and add a small portion of the mixture to it. (2 marks)

(2) Add a spatula full of the mixture to about a 6 cm depth of water in a boiling tube. Warm the mixture gently. Filter the contents of the boiling tube into another boiling tube and keep the filtrate. (1 mark)

(3) Divide the filtrate into four separate portions and place these in test tubes.

(a) To the first portion, add ten drops of aqueous silver nitrate. (2 marks)

(b) To the second portion, add ten drops of aqueous hydrogen peroxide.

(2 marks)

(c) To the third portion, add ten drops of dilute hydrochloric acid.

Use your observations to suggest what happened when the mixture was warmed with water in test (2). (2 marks)

(d) To the fourth portion, add ten drops of aqueous sodium carbonate. (1 mark)

Total: 10 marks

■ ■ ■

Candidate's response

Test	Observation	Inference
(1)	Carbon dioxide is given off	Mixture contains a carbonate
(2)	A white residue is obtained and there is a clear filtrate	
(3)(a)	A yellow precipitate is formed	There is an iodide present
(3)(b)	Brown solution is formed	
(3)(c)	No reaction	The carbonate has all reacted
(3)(d)	The solution clouds over	

Test **(1)** The candidate has made a common error. What is *observed* is 'bubbling' or 'effervescence'. No mark can be given for stating that the gas is carbon dioxide, even though in this case it is correct. The candidate also fails to record that the powder has (at least partly) dissolved to form a colourless solution. Therefore, no mark is awarded for the observation. Since the powder was a mixture of compounds, it is reasonable to suppose that a carbonate produced the bubbling during the reaction. So, 1 mark is gained for the suggested inference.

Test **(2)** The filtrate should be clear, but it is also colourless. The candidate has probably muddled the two words and by doing so loses the mark.

Test **(3)(a)** Both the observation and the conclusion are correct, for 2 marks.

Test **(3)(b)** It would have been better to write that the colourless solution turned brown, but the mark for the observation would probably be allowed. However, no inference has been made. A little thought should have reminded the candidate that iodine is brown in solution. The examiner was expecting that this observation would allow the candidate to confirm that an iodide was one of the components of the mixture.

Although this is a difficult mark, an answer should always be attempted.

Test **(3)(c)** The observation gains a mark, although it is not because the carbonate has reacted. The reason for the lack of bubbling is that the two components of the mixture have been separated and the filtrate does not contain the chemical that causes the bubbling.

Test **(3)(d)** 'Clouds over' is not an acceptable description for the formation of a precipitate. The candidate also fails to mention the white colour.

Overall, the mark of 5 is only grade-E standard, yet it would have been easy for this to have been raised to 8 or 9.

A2 task 1: reactions of a transition metal salt

For this assessment, candidates were given a dry sample of copper(II) chloride. They were not told what the sample was and it was simply labelled 'X'. They were asked to record their observations and decide what type of reaction was taking place when the following tests were carried out.

(1) Dissolve solid **X** in a minimum quantity of water in a boiling tube.　　(2 marks)

(2) Dilute the solution with more water. Keep the diluted solution for the tests that follow.　　(2 marks)

(3) Divide the solution obtained in part (2) into two portions.

　(a) To one portion add potassium iodide solution.　　(2 marks)

　(b) To the second portion:
　　(i) add aqueous ammonia dropwise until no further change occurs　(2 marks)
　　(ii) add dilute sulfuric acid to the solution obtained from part 3(b)(i)
　　　　(2 marks)

Total: 10 marks

■ ■ ■

Candidate's response

Test	Observation	Reaction type
(1)	Solid dissolves readily	Solubility
(2)	Solution goes blue	Dilution
(3)(a)	Brown solution forms with a white precipitate	Redox reaction
(3)(b)(i)	A dark-blue solution forms	Complex formation
(3)(b)(ii)	Solution goes lighter blue	Complex is broken down

This is quite a complicated sequence of reactions. The marking points are given in the table below:

24

Test	Observation	Reaction type	Mark
(1)	Dark-green solution	Complex ion formed	2
(2)	Solution goes blue	Ligand substitution/exchange	2
(3)(a)	Brown solution *and* a white precipitate form	Redox reaction	2
(3)(b)(i)	Light-blue precipitate Dark-blue solution	Precipitation Complex ion is formed	1 1
(3)(b)(ii)	Light-blue solution	Ligand substitution/exchange	2

Test **(1)** The candidate seems to have regarded the dissolving of the solid as not really being a reaction. At A2, it should be understood that solutions are formed by water binding to the cation to make a complex ion (as well as the anion being hydrated). The candidate also fails to record the green colour of the solution. No marks are awarded.

Test **(2)** The green solution contains a mixture of yellow $[CuCl]_4{}^{2-}$ ions and blue $[Cu(H_2O)_6]^{2+}$ ions. Water gradually replaces the chloride ion ligands. The examiner does not expect this process to be explained, but it should be understood that a colour change in a reaction involving a transition metal is an indication that a ligand substitution has taken place. The candidate gains 1 mark for the colour change.

Test **(3)(a)** The candidate has described and understood test **(3)(a)** well and gains both marks. The formation of iodine from a solution of potassium iodide is a sure indication that a redox reaction has taken place.

Test **(3)(b)(i)** The candidate has missed one of the observations. The aqueous ammonia was probably added too quickly and the first reaction, which can be easily overlooked, was not noticed. It is important that any instructions that are given are followed exactly. At first, copper(II) hydroxide is precipitated, which then dissolves to form the dark-blue complex ion, $[Cu(NH_3)_4(H_2O)_2]^{2+}$. The candidate scores 1 of the 2 marks.

Test **(3)(b)(ii)** The copper ammine complex ion is broken down and the $[Cu(H_2O)_6]^{2+}$ ion is formed again. This reaction therefore involves an exchange of ligands. The change in colour should have guided the student to provide the correct reaction type. The statement 'complex is broken down' is insufficient to earn the second mark.

The examiner would accept suitable alternative wording in the answers but, even so, this candidate scores only 5 marks out of 10.

A2 task 2: reactions of a double salt

The candidates were given a solid sample of ammonium iron(II) sulfate crystals. They were not told what the sample was and it was simply labelled 'Y'. They were asked to record their observations and conclusions when the following tests were carried out.

(1) Add a spatula full of solid Y to a boiling tube and add ten drops of aqueous sodium hydroxide. Warm the boiling tube. (2 marks)

(2) Place two spatulas full of solid **Y** into a boiling tube and add water to make a depth of approximately 6 cm in the tube. Warm the boiling tube gently to dissolve the solid completely. If necessary, add more water to make sure the solid is dissolved fully. Divide the solution between five separate test tubes and then perform the following tests.

(a) To the first portion, add ten drops of aqueous barium nitrate.   (1 mark)

(b) To the second portion, add ten drops of aqueous silver nitrate.   (2 marks)

(c) To the third portion, add ten drops of aqueous sodium hydroxide.   (1 mark)

(d) To the fourth portion, add ten drops of aqueous hydrogen peroxide followed by ten drops of aqueous sodium hydroxide.   (2 marks)

(e) To the fifth portion, add ten drops of aqueous sulfuric acid. Warm the tube gently and then add aqueous potassium manganate(VII) drop by drop.   (2 marks)

Total: 10 marks

■ ■ ■

Candidate's response

Test	Observation	Inference
(1)	Ammonia is given off and the substance melts	Substance contains ammonia and decomposes when heated
(2)(a)	White precipitate forms	This is a precipitation reaction
(2)(b)	No reaction	The substance is not a halide
(2)(c)	Green precipitate forms	The compound contains iron
(2)(d)	Solution goes more orange when hydrogen peroxide is added. Orange precipitate forms with the sodium hydroxide	The iron is oxidised
(2)(e)	Potassium manganate(VII) loses its colour	The compound can be oxidised

Test **(1)** Although this response is partially correct, it would not score any marks. The observation is a smell of ammonia and the substance dissolves into the water contained in the crystals. The word melt can only be used if a solid substance changes into a liquid. Most crystals will at least partially dissolve in their own waters of crystallisation when warmed so have a close look at a substance before it is heated to decide if this is a possibility. The inference is that the substance contains an ammonium ion, *not* that it contains ammonia.

Test **2(a)** The observation is correct and no further inference is possible, so the candidate gains the mark. (Some students might know that this is a test that probably identifies the anion as a sulfate. However, this is not in the specification and is not expected.)

Test **2(b)** The candidate should have looked at the 2 marks available for this reaction and been alerted that there must be more to it than has been noticed. In fact a grey/black precipitate of silver forms, although the reaction is rather slow and easy to miss. A redox reaction has taken place. The candidate fails to score.

Test **2(c)** The observation is straightforward and both it and the inference are required for the mark. The green precipitate indicates that Fe^{2+} ions are present. The vague statement that iron is present is insufficient for the mark.

Test **2(d)** The observations are correct and although the inference should state that Fe^{2+} has been oxidised to Fe^{3+}, the statement is just sufficient to allow both marks to be awarded.

Test **2(e)** The observation and inference are enough to earn the 2 marks.

Overall, the candidate scores 5 marks out of 10, which is just a pass.

Quantitative tasks

The quantitative task requires an experiment to be performed carefully and the results to be recorded accurately. The data obtained then have to be interpreted to calculate an unknown quantity or to come to some other valid conclusion. Although this might sound daunting, there are only a limited number of experiments that can be completed within the time limit allowed for the assessment. This means that it is possible to prepare for the task quite thoroughly so that the techniques and understanding required are well practised.

What knowledge is required?

AS and A2

There is little specific knowledge of reactions needed to complete the practical part of these tasks, as full instructions on how to carry out the experiment will be given. The emphasis is on the ability to interpret the results and carry out any related calculations. In general, calculations are broken down into small steps. However, it is necessary to understand each part of a calculation and to be confident in handling the mathematics required. The theory behind these calculations can be found in any standard A-level textbook.

Titrations

Many experiments that you do during your A-level course give only approximate results, but titrations should be accurate. It is essential that you are able to use volumetric flasks, pipettes and burettes confidently and reliably. The procedure to be followed when carrying out a titration is always the same and the correct techniques must be learned. These are explained in a later section (see pp. 33–34). The correct presentation of the results is also important and this is explained on p. 35.

To use the results to complete a calculation you need to be able to write balanced equations and understand how to interpret them. Equations may be supplied, but this must not be assumed. At AS, the equations concerned are those for the reactions of acids with hydroxides and carbonates (see p. 11). The calculations that follow a titration vary and the only way to be confident is to practise them until you become competent.

Enthalpy experiments

Enthalpy experiments are quite likely to appear as tasks. The experimental procedure is straightforward, requiring only the measurement of masses and volumes together with the temperature change that occurs as a result of a reaction. The calculation requires you to know how to determine the amount of heat released per gram and then per mole. It is likely that you will then need to understand how to use Hess's law to deduce the enthalpy change for an unknown reaction.

Water of crystallisation

You may have to carry out an experiment that involves heating crystals of a substance to drive off the water of crystallisation. This is an experiment that only gives good results if it is carried out very carefully. Take great care not to heat too strongly or the crystals may 'spit' and jump out of the container. There is also a danger that the anhydrous product may decompose if you overheat the sample. Remember that you should not weigh a hot container on a balance. From the results, the percentage by mass of water of crystallisation present can be found and, from that, the formula of the crystals.

Measuring the volume of a gas

You should be familiar with collecting a gas either by displacement of water from a measuring cylinder or by using a syringe. It is easy when carrying out experiments involving gases to allow some of the gas to escape, so you should make sure that you practise experiments of this type. You should be ready to plot graphs and do calculations based on the measurements taken.

A2 only

Rate experiments

A rate experiment is a likely task that may also be examined as part of an evaluative exercise. It will almost certainly require you to plot an appropriate graph. There are different ways of carrying out rate experiments and interpreting the data obtained and these must not be confused.

It is most likely that the experiment will be an 'initial rate' procedure where an experiment is repeated several times using different concentrations of reactants. The time, t, is measured to a fixed point in the reaction. For example, this might be to when a fixed amount of precipitate has formed or a particular depth of colour is obtained. In this case the graph that should be plotted is rate against concentration, where rate is measured as $1/t$. The assumptions behind the use of $1/t$ are discussed on page pp. 66–67.

It is also possible that the experiment could involve making a mixture and then allowing the reaction to run its course with the measurement of some quantity (such as a reduction in the concentration of a reactant) being recorded at suitable time intervals. In this case, a graph of concentration against time can be plotted and the rate determined by taking the gradient at various points. A zero-order reaction can be recognised as the graph will be a straight line and a first-order reaction can be identified by noting that the half-life is constant.

Electrode potentials and redox reactions

Electrode potentials are difficult to measure reliably in the laboratory, so it is unlikely that you will be asked to do this. (It could, however, be part of an evaluative task.) The use of electrode potentials to determine whether a reaction is feasible might be included in a quantitative task. You could be asked, for example, to perform various redox reactions and support your conclusions using relevant redox potentials. This could be followed by asking for the equation for the reaction to be deduced using the half-equations involved. This is an area where a good understanding of the theory behind electrode potential is required, so this must be revised carefully before the task is attempted.

Acids and bases and equilibria

The reactions of acids and bases could be asked, although this would be most likely within a qualitative task or a titration or maybe as part of an enthalpy experiment. pH meters may have been used in class experiments but these will not feature in an assessment. However, you should understand how the strength of acids and bases affects their degree of ionisation in aqueous solution and their likely reactivity.

The calculation of an equilibrium constant is an unlikely possibility because equilibria are very slow to establish. Nevertheless, the effect of concentration and temperature on the position of an equilibrium should be appreciated and could be relevant to the interpretation of some results.

What skills are required?

The quantitative tasks will normally be based on skills that have been well practised. Some of the measurements required are straightforward and you should be comfortable with the use of balances, thermometers and measuring cylinders. The importance of the correct use of volumetric equipment (pipettes, burettes and volumetric flasks) cannot be over-estimated. There will always be marks awarded for the achievement of accurate results in a titration and probably in an enthalpy experiment.

Balances

These are simple to use but most are very sensitive. Make sure that the pan of the balance is clean before using it and do not attempt to weigh anything that is much warmer than room temperature. Warm currents of air may make the reading unsteady and unreliable. Be careful to read any instructions carefully — for example, you may usually use the tare on the balance when weighing a solid but, in a task, you should check whether this is expected or not.

Non-digital thermometers

If you are reading the temperature of a solution, make sure the bulb is fully covered by the liquid. As with all measuring equipment, be careful to take readings with your eyes at the same level as the top of the liquid. Do not view it at an angle because this makes it easy to under- or over-estimate the measurement.

Measuring cylinders

You should appreciate that measuring cylinders have limited accuracy (see p. 34) and so are not appropriate for the most accurate work. In the instructions for a task they will only be specified when a precise volume is not needed.

practical tasks

Using a pipette

Used correctly, a pipette delivers a precise volume (see Figure 1). Since its use requires sucking liquid into the pipette, a safety bulb must always be used.

When the pipette is filled, the meniscus of the liquid should sit on the volume mark on the neck of the pipette. To do this reliably requires practice.

The solution in the pipette should be allowed to run out freely. A small amount of solution will remain in the bottom of the pipette. Touch the tip of the pipette on the surface of the liquid that has been run out and then ignore any further solution that remains in the pipette.

Figure 1 Using a pipette

Pipettes are available in a variety of sizes — $10.0\,cm^3$ and $25.0\,cm^3$ are frequently used. A $25.0\,cm^3$ pipette is usually capable of measuring a volume of liquid with a maximum error of $\pm0.06\,cm^3$, although it does depend on the quality of the manufacture. Some pipettes are graduated to allow you to dispense a volume less than the full capacity. These are useful but may be slightly less accurate than a standard pipette.

Using a burette

Like the pipette, when used correctly, the burette is very accurate. When filling the burette, first make sure that the tap is closed, and then use a funnel so that you don't spill solution down the outside. Be careful not to overfill the burette. Now remove the funnel. Run some solution out so that the space below the tap becomes filled.

When using a burette, the volume used is measured as the difference between the readings made at the start and at the finish. It is usual to read from the bottom of the meniscus (see Figure 2 on p. 32) but, because you are subtracting two readings, you could use the top of the solution. This may be helpful if the solution in the burette is coloured — for example, at A2, when using potassium manganate(VII) for a redox titration.

Figure 2 Reading a burette

Burettes usually have gradations at every $0.1\,cm^3$ and any reading has a maximum error of $\pm 0.05\,cm^3$. However, they do come in different grades so it is worth checking the level of accuracy of the type you are using. The volume measured by a burette cannot usually exceed $50.0\,cm^3$.

Using a volumetric flask

As with all volumetric equipment, the volumetric flask is very accurate. The flask holds the specified volume when it is filled so that the meniscus of the solution sits on the mark on the neck. If you are dissolving a solid in water to make a solution some care is needed. It is usual to dissolve the solid first in a container such as a beaker containing some distilled water. This allows you to warm the solution if necessary to encourage the solid to dissolve. Once you have dissolved the solid and, if necessary, allowed the solution to cool back to room temperature you can transfer the solution to the volumetric flask, pouring it in through a funnel. Wash any remaining solution from the beaker into the flask and be careful not to lose any solution while doing this. You should now add distilled water to the flask until the meniscus is sitting on the mark. Finally, invert the flask a few times (with the stopper securely in place) to ensure that the solution is mixed completely. *Never* warm a volumetric flask. First, it is not constructed to be heated and it will therefore break readily. Second, even if the flask survives the heating, the volume has been calibrated by the manufacturer to be correct only at room temperature.

A volumetric flask may be needed if a solution has to be accurately diluted. For example, $10.0\,cm^3$ of a $1.00\,mol\,dm^{-3}$ solution could be measured using a pipette and then transferred to a $250\,cm^3$ volumetric flask. If water is then added to fill the volumetric flask until the meniscus of the solution sits on the mark, the solution will have been diluted to a concentration of $\left(\dfrac{10.0}{250}\right) \times 1.00 = 0.0400\,mol\,dm^{-3}$.

Titrations

A titration that is carried out carefully can be very accurate. In most cases, a volume can be measured consistently to an accuracy of $\pm0.1\,cm^3$. However, this does require you to master the techniques involved. Only practice in the laboratory will help you achieve this, but there are a number of points that can be made.

- The solution to be measured in the pipette should be placed in a conical flask rather than in a beaker because it is easier to shake a conical flask without spilling the solution.
- Don't forget to remove the funnel from the top of the burette before taking the initial reading. Sometimes drops remaining in the funnel can fall into the burette while the titration is being carried out.
- All acid–base titrations (and most others) require an indicator to show when the reaction between the solution in the conical flask and the solution added from the burette is complete. Do not add more than the minimum amount of indicator that is necessary to see the colour change at the end point of the reaction. Your eye is more sensitive when the colour changes are not too intense.
- During the titration make sure that the solution being added is mixed thoroughly with the solution in the conical flask. This requires gentle swirling, ideally throughout the course of the titration, but certainly when close to the end point of the reaction.
- Some of the solution from the burette may fall down the inside of the conical flask, rather than into the solution it contains. The addition has been measured by the burette, so you must make sure that all the solution added has a chance to react. Sometimes, you can achieve this by swirling the solutions together but water can also be squirted down the side of the flask to wash everything into the solution below. This is often a source of concern to students who believe that this will lead to an incorrect value for the titration. This is not so. Although the solution is undoubtedly diluted, the amount, in moles, present in the pipetted solution will not have been changed and it is this that is being titrated.
- Near the end point, the burette must be used particularly carefully because one drop of solution will eventually change the colour of the indicator. This means that the tap of the burette must be controlled exactly. By turning the tap very slowly to the correct position, it is perfectly possible to deliver the solution drop by drop. That way you can stop the addition between drops if necessary. Do not be tempted to add

the solution by turning the tap on and off quickly in the hope that only one drop will be added at a time. Instead, you will get small bursts of solution that are certainly greater than a single drop.

Recording results

Each piece of apparatus that is used in the laboratory to measure a quantity has an accuracy that depends on the quality of its manufacture. For example, a basic balance may only give a measurement to the nearest gram but, with better equipment, a reading to one, two, three or even four decimal places can be made. The volume of a liquid can be measured approximately using a beaker with lines marked on the side, but a measuring cylinder is more accurate and volumetric apparatus is better still. For most measuring equipment the manufacturer will have stated the accuracy that it possesses and sometimes this is etched onto the apparatus.

When recording a measurement, the intended precision should be indicated using the appropriate number of decimal places. A mass written as $31.21\,g$ indicates that the measurement has an accuracy close to the value of the last digit quoted (i.e. to the nearest $0.01\,g$). It is important to realise that the way a measurement is recorded implies the precision that is intended. This is usually clear in the case of a balance, but the same principle must be applied to apparatus where it may be less obvious. The scale on the side of a measuring cylinder does not give any indication of the level of accuracy that can be expected for a measurement made using it. The only way to deal with this is to find out what the manufacturer quotes for this particular piece of apparatus. If the accuracy is to only $\pm 1\,cm^3$ (which is quite likely for a $100\,cm^3$ measuring cylinder) then all readings must be recorded only to the nearest whole number. It would be wrong to write $75.0\,cm^3$ when it should be $75\,cm^3$.

There is a general rule of thumb for non-digital apparatus that the accuracy can be taken to be half a division of the marked scale. This can be helpful if no further guidance is available, but it should be understood that it is no more than a broad indication and does not always apply.

The following is a guide to the apparatus you are most likely to use:

- Record results obtained from digital equipment showing all the decimal places displayed. This applies, for example, to a balance. For example, $24\,g$ on a two-decimal-place balance must be written down as $24.00\,g$. Be careful to give all readings to the same number of decimal places.
- The accuracy of measuring cylinders varies considerably and you should check whether you have been given any specific information about the equipment you are using. If not, the rule of thumb mentioned above is useful and the maximum accuracy should be considered to correspond to half the distance between the markings. For example, a $250\,cm^3$ cylinder might be marked in divisions of $2\,cm^3$. In which case, the maximum accuracy can be taken to be to the nearest $1\,cm^3$. Therefore, it would be inappropriate to include any decimal points when recording the volume measured.

- Thermometers also vary in accuracy. No general rule other than the use of half a division can be given.
- A volumetric flask when filled carefully so that the meniscus is on the line usually gives a reading with a maximum accuracy of $\pm 0.2\,cm^3$ (or sometimes $\pm 0.3\,cm^3$). The volume should be quoted to one decimal place, for example $250.0\,cm^3$.
- The volume in a pipette (normally accurate to $0.06\,cm^3$) should also be given to one decimal place — for example, $25.0\,cm^3$.
- Burettes are normally accurate to $\pm 0.05\,cm^3$, so this is the maximum precision that should be recorded. Remember that this precision also applies to the zero, which should be recorded as $0.00\,cm^3$. Values such as $26.25\,cm^3$, $26.00\,cm^3$ and $26.05\,cm^3$ are all valid readings. Values such as $24.0\,cm^3$ (only one decimal place), $24.13\,cm^3$ (a greater level of precision than is justified) are not. When a titration has been completed the mean titre is required. Since the volume added from the burette is obtained by taking an initial and a final reading, both of which are only accurate to $\pm 0.05\,cm^3$, the mean is arguably only accurate to $\pm 0.1\,cm^3$ and you may, therefore, be asked to give this to one decimal place only.

Calculating the mean titre

When working out the mean titre, you should use only the concordant titres (i.e. those that agree most closely). If you have two titres that either agree exactly or are closer together than other volumes, then this is straightforward. However, sometimes you may need to use some judgement to select the appropriate titres. Some examples are given below.

Volumes added from the burette/cm^3	Titres to use	Mean titre/cm^3 (to 1 d.p.)
24.10; 24.20; 24.30	All three	24.2
23.55; 23.75; 23.95	All three and the mean 23.75 written to one decimal place as 23.8	23.8
26.20; 26.40; 26.70	Ignore 26.70 because the two readings 26.20 and 26.40 are close	26.3

Since the accuracy of equipment can vary, if questions are asked in an assessment then the precision you should use will be indicated — for example, 'the volume was measured using a volumetric flask with a maximum error of $\pm 0.2\,cm^3$' or 'the volume was measured using a burette with a maximum error in each reading of $\pm 0.05\,cm^3$'. The measurements can then also be used in an estimation of percentage error (see pp. 56–57).

Significant figures

The number of decimal places indicates the precision intended for any piece of apparatus being used. In quoting an answer during the course of a calculation, this information is given by providing the answer to an appropriate number of significant figures. For example, if a relative molecular mass has been calculated as 126.23 but the accuracy of the measurements does not allow for this to be quoted to more than three significant figures, the relative molecular mass should be given as the whole number, 126.

In simple cases, the number of significant figures is the number of digits in the answer. For example, 31.21 has four significant figures, while 31.2 has three significant figures and 31 has just two.

In some cases, numbers need rounding up or down before the answer is quoted to a particular number of significant figures. For example, 17.87 has four significant figures. To three significant figures, this is 17.9 (as 17.87 is nearer to 17.9 than to 17.8). To two significant figures, it is 18 (17.87 is nearer to 18 than to 17).

A number ending in '5' is always raised to the higher number. So, for example, 32.5 is written as 33 when quoted to two significant figures and 25.15 is written to 25.2 when quoted to three significant figures.

When a number less than 1 has zeros *after* the decimal point these are not considered to be significant. So, for example, 0.004 has only one significant figure because the zeros don't count and 0.0526 has three significant figures, which to two significant figures is 0.053.

A problem arises if a whole number ends with zeros. For example, it is not clear whether 1800 has four, three or two significant figures. In order to get round this difficulty standard index form (sometimes just called standard form) is used. To quote 1800 to two significant figures, you must write 1.8×10^{-2}. If four significant figures are intended, then it is written as 1.800×10^{-2}.

In the tasks, it will always be made clear if an answer must be given to a particular number of significant figures. It is a common mistake to forget that you have been asked to do this, so highlight it on the question paper so that you remember. If it is not asked for specifically then the examiner will allow a reasonable amount of leeway in your answer. Nevertheless, you should be sensible in the number of significant figures that you use. Consider the accuracy of the apparatus and don't quote your answer to a number of significant figures greater than this allows.

For example, in an enthalpy experiment, if a mass of 1.06 g of sodium carbonate reacted with 50.0 g of excess acid and a temperature rise of 3.9°C was measured, then, taking $4.18 \text{J} \text{g}^{-1} \text{K}^{-1}$ as the specific heat capacity of the solution, the heat produced could be calculated as:

$$q = 50.0 \times 4.18 \times 3.9 = 815.1 \text{ joules or } 0.8151 \text{ kJ}$$

But, since the temperature rise was only measured to an accuracy of two significant figures, the answer should also be quoted to two significant figures i.e. 0.82 kJ.

Significant figures during a calculation

To help you, any calculation required by a task will usually be broken down into several parts, each of which requires an individual answer. Where this is the case, you may be told to give only the *final* answer to an appropriate number of significant figures. You will not normally lose marks for using an inappropriate number of significant figures for the intermediate answers. However, this is not ideal and there is a better way of handling the situation:

- Give your answers to the various parts of the question to the number of significant figures that corresponds to the accuracy of the equipment.
- Use all the figures given by your calculator as an answer to one step to obtain the answer to the following step. If you use answers that have been rounded at each individual step to obtain the answer to the steps that follow, you will steadily lose the appropriate precision for your final answer.

Significant figures

(1) Write the number 15.2548 to:

    (a) five significant figures

    (b) four significant figures

    (c) three significant figures

(2) Write the number 0.021768 to:

    (a) four significant figures

    (b) three significant figures

    (c) two significant figures

(3) Write the number 0.01899 to:

    (a) three significant figures

    (b) two significant figures

(4) Using the appropriate number of significant figures write down:

    (a) a mass of exactly 23 g of solid weighed on a two-decimal place balance

    (b) 25 cm^3 of a solution measured in a pipette with a maximum error of $\pm0.06\,cm^3$.

(5) A student carries out a titration using a burette accurate to $\pm0.05\,cm^3$ and fills in a results table as shown below.

	Trial	1	2	3
Final burette reading/cm^3	24.4	24.02	23.8	24.06
Initial burette reading/cm^3	0	0	0	0
Titre/cm^3	24.4	24.04	23.8	24.06
Titres used to calculate the mean/cm^3		✓		✓
Mean titre value to 1 d.p./cm^3	24.05			

Write out the table correcting the errors that have been made by the student.

Answers to these questions are given on p. 88.

Using the results of an experiment

Full instructions will be given as to how the results of experiments should be used and you will have probably carried out similar experiments in class. However, before attempting the assessment make sure that you fully *understand* the background theory as you will not necessarily be asked to use the results in a familiar way.

Titration calculations

As these are certain to be required in at least one of the assessed practicals at AS or A2, a brief summary of the steps involved is given here. Any standard textbook will provide more detail. The basic procedure is the same whether it is an acid–base titration or, at A2, a redox titration.

(1) Obtain the mean titre from your burette readings.

(2) For one of the solutions, the concentration in $mol\,dm^{-3}$ is normally known. Use the mean volume obtained from the titration to calculate the amount in moles, n, present in this volume.

(3) Refer to the balanced equation to determine the amount in moles that will react with n. This is the amount in moles contained in the second solution used in the titration.

(4) Convert the information in **(3)** to obtain the concentration of the second solution in $mol\,dm^{-3}$.

These four steps will not allow you to solve everything that might be asked, but titration calculations should be practised until you are completely familiar with this procedure.

Plotting graphs

If you are lucky, the axes needed to plot a graph will be supplied. If not, then the choice of axes is important and will be likely to carry marks. You should make sure that, as far as possible, the full size of the graph paper is used. Be careful however not to choose a scale that is so awkward that errors are made when plotting the points. Always check that each axis is labelled and include the units.

Other calculations

Other exercises may involve calculations based on chemical formulae, such as calculating the percentage of water of crystallisation present in crystals. The relationship between the volume of a gas and the amount in moles could be asked and an understanding of the information contained in a balanced equation forms the basis of this type of calculation.

At A2, practicals involving electrode potentials are possible, but this largely involves an understanding of their use rather than performing a calculation. Calculations of pH and perhaps the value of an equilibrium constant might be expected. If you are advised, in advance, that a particular topic is the focus of a task then obviously the work relating to this should be revised.

Exemplar quantitative tasks

The interpretation of the results obtained from a quantitative task is important and carries a reasonable proportion of the marks. The proper use of apparatus should have been practised sufficiently for all the marks available for carrying out the task to be scored. There may be as many as 9 out of 15 marks for carrying out and recording the results of a titration before the calculation is started.

The most significant of these marks are for obtaining results from the experiment that agree with your teacher's value. The mark will take account of the accuracy of the equipment used, so the agreement does not have to be perfect. For example, to obtain full marks, the mean titre obtained from a titration might have to agree within $0.2\,cm^3$. Other values would gain some credit so long as they were not too far out.

To obtain the marks for recording the results you have to provide all measurements to the correct number of decimal places and do simple things such as subtracting numbers correctly.

In these examples of quantitative tasks, the layout is broadly similar to that used in the assessments, but you should expect their layout to be rather more formal. You would, for example, expect the colour changes of indicators to be given and other details to be more precise.

AS task 1: to calculate the mass of the ion X^- in 1 mole of an acid HX

You are provided with sodium carbonate crystals, $Na_2CO_3.10H_2O$ and a solution of the acid, HX. You will use a titration to determine the concentration of the HX solution and hence deduce the mass of 1 mole of X^-.

All readings should be recorded in the tables given.

Procedure:

(1) Weigh the empty container provided.

(2) Add between 3 g and 4 g of sodium carbonate crystals, $Na_2CO_3.10H_2O$, to the container and reweigh it.

(3) Dissolve the crystals in $150\,cm^3$ of water in a beaker.

(4) Pour the solution into a $250\,cm^3$ volumetric flask and make up the solution to $250.0\,cm^3$

(5) Fill a burette with the acid HX.

(6) Use a pipette to take $25.0\,cm^3$ of the sodium carbonate solution and place this in a conical flask.

(7) Add 3 drops of methyl orange indicator.

(8) Titrate the sodium carbonate solution against the acid HX.

(9) Repeat the titration to obtain two consistent results.

Procedure total: 9 marks

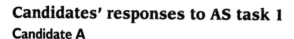

Candidates' responses to AS task 1
Candidate A

Mass of container + sodium carbonate crystals/g	29.92
Mass of container/g	26.72
Mass of sodium carbonate crystals/g	3.2

	Trial	1	2	3
Final burette reading/cm³	24.40	23.90	23.60	24.20
Initial burette reading/cm³	50	50	50	50
Titre/cm³	24.40	23.90	23.60	24.20
Titres used to calculate the mean/cm³		✓	✓	✓
Mean titre value to 1 d.p./cm³	23.9			

Candidate B

Mass of container + sodium carbonate crystals/g	31.12
Mass of container/g	27.45
Mass of sodium carbonate crystals/g	3.67

	Trial	1	2
Final burette reading/cm³	25.70	25.40	25.35
Initial burette reading/cm³	0.00	0.00	0.00
Titre/cm³	25.70	25.40	25.35
Titres used to calculate the mean/cm³		✓	✓
Mean titre value to 1 d.p./cm³	25.4		

Candidate A has written 3.2 g instead of 3.20 g and so loses the mark for recording the mass used. All the results must be recorded to a consistent number of decimal places. Candidate B scores the mark.

There are 2 marks for recording the titration results. The first is for the completion of the table with all the burette readings included and the second is for using a consistent number of decimal places for the readings. Candidate A fails to score these marks. The initial reading is 0.00 cm³, not 50.00 cm³. It is the reading on the burette that should be recorded, not the volume of solution in the burette. (Note that when the solution is at the zero mark, the burette contains more than 50.00 cm³ because there is some solution in the bottom of the burette below where the gradations begin.) Candidate A also fails to be consistent in the use of two decimal places by writing 50, rather than 50.00. Candidate B scores both marks.

There are 2 marks for obtaining consistent results. Candidate A has not carried out the titration with sufficient care and has obtained results that differ by 0.30 cm³, so scores only 1 mark. Candidate B has been efficient and careful and obtains 2 marks.

There is 1 mark for obtaining the correct mean value and both candidates score this mark.

There are 3 marks available for how closely the results obtained by the candidates compare with the teacher's result. The different masses used by the students are taken into account. In this case the teacher, like Candidate A, used 3.20 g but obtained a mean titre of 23.6 cm³. Candidate A's result of 23.9 cm³ is not sufficiently accurate for all 3 marks, but 2 marks are allowed as the value is reasonably close. Candidate B's titration result is close enough for 3 marks. (Note that the accuracy expected for the candidates' titration results varies according to the difficulty of the titration and will not necessarily be as described in this case.)

Overall, for carrying out the experiment, Candidate A scores only 4 marks out of 9; Candidate B scores all 9 marks.

Analysis:

(a) **Calculate the mass of 1 mol of sodium carbonate crystals.** (1 mark)

(b) **Calculate the amount, in moles, of sodium carbonate that you used to make up the 250.0 cm³ of solution.** (1 mark)

(c) **Calculate the amount, in moles, contained in the 25.0 cm³ of the sodium carbonate solution used for the titration.** (1 mark)

The equation for the reaction taking place during the titration is:

$$Na_2CO_3(aq) + 2HCl(aq) \rightarrow 2NaCl(aq) + CO_2(g) + H_2O(l)$$

(d) **Use the equation and your answer to part (b) to state the amount, in moles, of HX that was present in the mean titre volume of HX.** (1 mark)

(e) **Calculate the concentration, in mol dm⁻³, of the HX solution.** (1 mark)

(f) **If the HX solution has a concentration of 3.45 g dm⁻³, calculate the mass of 1 mole of X⁻.** (1 mark)

Analysis total: 6 marks

■ ■ ■

Candidates' response

Candidate A

(a) mass of $Na_2CO_3.10H_2O = 286.0\,g$

(b) amount in moles is $\dfrac{3.2}{286} = 0.01119\,mol$

(c) concentration in $mol\,dm^{-3}$ is $4 \times 0.0.1119 = 0.044755\,mol\,dm^{-3}$

In $25.0\,cm^3 = \dfrac{0.044755}{40} = 0.001119\,mol$

(d) mol of acid $= 2 \times 0.001119 = 0.002238\,mol$

(e) concentration of acid $= 0.002238 \times \left(\dfrac{1000}{23.9}\right) = 0.09363\,mol\,dm^{-3}$

(f) $3.45\,g = 0.09363\,mol$

HX has mass of 36.847 and $1\,mol$ of X has a mass of $35.847\,g$.

Candidate B

(a) mass of $1\,mol$ of $Na_2CO_3 = 106\,g$

(b) mol used was $\dfrac{3.77}{106} = 0.0356\,mol$

(c) 0.0356 is what I used in my experiment

(d) The mol of HX was twice as much $= 0.0712\,mol$

(e) In a dm^{-3} there are $0.0712 \times \dfrac{1000}{25.4} = 2.803\,mol\,dm^{-3}$

(f) The acid is $3.45 \times 2.803 = 9.67\,g$, so X is $8.67\,g$.

🖉 Candidate A is confident with titration calculations and completes the calculation correctly. However, part (c) was not done in a sensible way. The answer to part (b) was the amount in moles in $250.0\,cm^3$ so the amount, in moles, in $25.0\,cm^3$ is the answer to (b) divided by 10. The candidate has obtained the correct answer and will, therefore, not lose any marks. The questions do not specify that the answers must be given to a particular number of significant figures, but it would have been more sensible to quote the final answer to no more than four significant figures. Nevertheless, Candidate A scores all 6 marks.

Candidate B is more uncertain and makes mistakes in the calculation. The mass calculated in part (a) is incorrect. $106\,g$ is correct for the formula Na_2CO_3 but it was the crystals of formula $Na_2CO_3.10H_2O$ that were weighed out. Part (b) is therefore also wrong but, as the student has carried it out correctly, the mark is awarded. This is referred to as an 'error carried forward' and is allowed in calculations so that a mistake early on does not mean that all the marks for the subsequent steps are lost. The candidate does not understand what has been asked in part (c) and fails to score. Parts (d) and (e) have both been carried out in the right way so, despite the fact that the figures are wrong, 2 marks are awarded. The candidate does not understand part (f) and fails to score. Candidate B therefore obtains 3 marks for the calculation.

Overall, Candidate A scores 10 marks out of 15 and Candidate B scores 12. Candidate A is the more able of the two students but Candidate B, by concentrating on doing the experiment well, has obtained the better mark. By carrying out the experiment more carefully, Candidate A could have obtained 14 or 15 marks.

AS task 2: to determine the enthalpy of crystallisation of magnesium sulfate

You are provided with solid anhydrous magnesium sulfate, $MgSO_4$, and magnesium sulfate crystals, $MgSO_4.7H_2O$. You will determine the enthalpy change when these are dissolved separately in excess water. You will use the results to calculate the enthalpy change of the reaction:

$MgSO_4(s) + 7H_2O(l) \rightarrow MgSO_4.7H_2O(s)$

Record all your results in the tables provided.

Procedure:

(1) Use a measuring cylinder to pour $100\,cm^3$ of water into a plastic cup.

(2) Measure the temperature of the water to the nearest 0.5°C.

(3) Weigh the bottle containing the anhydrous magnesium sulfate.

(4) Pour the anhydrous magnesium sulfate into the water. Stir vigorously using the thermometer and record the highest temperature that is reached to the nearest 0.5°C.

(5) Weigh the empty weighing bottle.

(6) Repeat the experiment using the magnesium sulfate crystals and record the lowest temperature that is reached.

Procedure total: 7 marks

■ ■ ■

Candidates' responses to AS task 2
Candidate A
Experiment 1

Mass of container + anhydrous magnesium sulfate/g	32.34
Mass of container/g	24.23
Mass of anhydrous magnesium sulfate/g	8.11

Final temperature/°C	27.0
Initial temperature/°C	19.5
Temperature rise/°C	7.5

Experiment 2

Mass of container + magnesium sulfate crystals/g	32.60
Mass of container/g	24.37
Mass of magnesium sulfate crystals/g	8.23

Final temperature/°C	19.0
Initial temperature/°C	20.0
Temperature fall/°C	−1

Candidate B

Experiment 1

Mass of container + anhydrous magnesium sulfate/g	33.65
Mass of container/g	25.75
Mass of anhydrous magnesium sulfate/g	7.9

Final temperature/°C	25.5
Initial temperature/°C	19
Temperature rise/°C	6.5

Experiment 2

Mass of container + magnesium sulfate crystals/g	33.86
Mass of container/g	25.70
Mass of magnesium sulfate crystals/g	8.16

Final temperature/°C	18.3
Initial temperature/°C	19.5
Temperature fall/°C	−1.2

The 7 marks are for the correct presentation of the results and agreement with the teacher's values for the two experiments.

Candidate A makes just one mistake in the presentation of the results; the temperature fall in the second experiment should be recorded as 1.0, rather than 1. This costs the candidate 1 mark. There is no need for the minus sign before 1.0 because the table indicates that there is a temperature fall. However this would not result in the loss of a mark. In both experiments, Candidate A's results agree closely with the teacher's results and the candidate scores the 3 marks available for this.

Candidate B is more careless. In the first experiment, there are two instances of figures being recorded incorrectly. The mass of anhydrous magnesium sulfate should be given as 7.90 g and the initial temperature should be 19.0°C. The candidate loses 1 mark. In the second experiment, the thermometer has been read to a greater accuracy than the 0.5°C specified. The figure of

18.3°C loses a mark. In addition, the result of the first experiment does not agree closely with the teacher's value and only 2 marks out of 3 are awarded.

Overall, for carrying out the experiment, Candidate A scores 6 marks and Candidate B scores only 4 marks.

Analysis:

The specific heat capacity of the solution of magnesium sulfate is $4.2\,J\,g^{-1}\,K^{-1}$.

(a) **Use the results of the first experiment to answer the questions that follow.**

    (i) **Calculate the heat produced as the anhydrous magnesium sulfate dissolves.**

                                                      (1 mark)

    (ii) **Calculate the enthalpy change, in $kJ\,mol^{-1}$, for this reaction.**   (2 marks)

(b) **Use the results of the second experiment to answer the questions that follow.**

    (i) **Calculate the heat produced as the magnesium sulfate crystals dissolve.**

                                                      (1 mark)

    (ii) **Calculate the enthalpy change, in $kJ\,mol^{-1}$, for this reaction.**   (2 marks)

(c) **Calculate the enthalpy change for the reaction:**

    $MgSO_4(s) + aq \rightarrow MgSO_4.7H_2O(s)$

    Give your answer to three significant figures.   (2 marks)

                                              Analysis total: 8 marks

■ ■ ■

Candidates' responses

Candidate A

(a) (i) heat produced is $\dfrac{(108.11 \times 4.2 \times 7.5)}{1000} = 3.4055\,Kj$

    (ii) r.m.m. of $MgSO_4 = 24.3 + 32.1 + 64.0 = 120.4$

        $\Delta H = 3.4055 \times \left(\dfrac{120.4}{8.11}\right) = -50.56\,Kj\,mol^{-1}$

(b) (i) Heat lost $= \dfrac{(108.23 \times 4.2 \times 1)}{1000} = 0.455\,Kj$

    r.m.m. of $MgSO_4$ crystals $= 120.4 + 7 \times 18.0 = 246.4$

    (ii) $\Delta H = 0.455 \times \left(\dfrac{246.4}{8.23}\right) = +13.6\,Kj\,mol^{-1}$

(c)

$$MgSO_4(s) + 7H_2O(l) \xrightarrow{\Delta H} MgSO_4.7H_2O(s)$$

$\Delta H_a \searrow$ $\nearrow \Delta H_b$

       $MgSO_4(aq)$

$\Delta H + \Delta H_b = \Delta H_a$

$\Delta H = \Delta H_a - \Delta H_b = -50.56 - 13.6 = -64.16\,Kj\,mol^{-1}$

Candidate B

(a) (i) $100 \times 4.2 \times 6.5 = 2730 \, J$

(ii) r.m.m. of $MgSO_4 = 24 + 32 + 64 = 120$

enthalpy change $= 2730 \times \left(\dfrac{120}{7.9}\right) = 41468.4 \, J = 41.4 \, kJ \, mol^{-1}$

(b) (i) $100 \times 4.2 \times 0.8 = 336 \, J$

r.m.m. of $MgSO_4.7H_2O = 120 + 7 \times 18 = 246$

(ii) enthalpy change $= 336 \times \left(\dfrac{246}{8.16}\right) = 10129.4 \, J = 101.3 \, kJ \, mol^{-1}$

(c) change for $MgSO_4 + 7H_2O \rightarrow MgSO_4.7H_2O$

$= 41.4 - 101.3 = 59.9 \, kJ \, mol^{-1}$

Candidate A is confident with the calculation. In calculating the heat produced, it is acceptable to take the mass as either 100 g, which is just the mass of the water, or to do as A has done and use 100 g plus the mass of the dissolved solid. However, the candidate makes two mistakes. The first is writing the units as $Kj \, mol^{-1}$, rather than $kJ \, mol^{-1}$. This is an error that is seen often. The units are usually, but not always, supplied on the paper, so it is important to learn them. As the units are very nearly correct, the examiner would probably not deduct a mark, but this may not always be the case. However, 1 mark is lost because Candidate A has forgotten that the final answer in part (c) must be quoted to three significant figures. The candidate is rather careless throughout the calculation in the use of significant figures. This might matter in some calculations though the examiner will allow some leeway. Candidate A scores 7 marks out of 8 for the interpretation of the results.

Candidate B is less successful. The answer to part (a)(i) is correct. If the paper does not specify the units required, an answer in either joules or kilojoules is accepted. Both marks are lost in part (a)(ii). The first is lost for not using the atomic mass values in the periodic table provided, which are to one decimal place. Another mark is lost because Candidate B has forgotten to put a minus sign before the figure 41.4. As the reaction is exothermic, this *must* be included. The figure is also rounded incorrectly to 41.4 rather than 41.5. The candidate gains the marks for (b)(i) and (b)(ii). The error in the atomic masses is repeated but would not be penalised a second time and, in this experiment, the reaction is endothermic so ΔH is positive. Candidate A carefully drew an enthalpy cycle and obtained the correct answer to part (c), but Candidate B has tried to get the answer without the help of the cycle. In doing so a mistake is made in the calculation and so 1 mark is lost. However, the answer is quoted to three significant figures, so the candidate gains this mark.

Overall, Candidate A scores 13 marks out of 15, which is a good mark. Candidate B has made many careless mistakes and scores only 9 marks.

A2 task 1: analysis of copper sulfate crystals

You are provided with crystals of copper sulfate, $CuSO_4.xH_2O$, that have been prepared by reacting 5.00 g of copper oxide with excess sulfuric acid.

You will first use a titration to determine the value of x.

You will then calculate the percentage yield that was obtained in the reaction between copper oxide and sulfuric acid.

All readings should be recorded in the tables provided.

Procedure:

(1)   Weigh the empty container provided.

(2)   Add the sample of copper sulfate crystals to the container and reweigh it.

(3)   Dissolve the crystals in 150 cm³ of water in a beaker.

(4)   Pour the solution into a 250 cm³ volumetric flask and make up the solution to 250.0 cm³.

(5)   Fill a burette with a solution of 0.100 mol dm⁻³ sodium thiosulfate.

(6)   Use a pipette to take 25.0 cm³ of the copper sulfate solution and place this in a conical flask.

(7)   Use a measuring cylinder to add approximately 10 cm³ of the potassium iodide solution provided. This is an excess of potassium iodide.

(8)   A reaction takes place between the copper ions and the iodide ions. Copper(I) iodide is precipitated and iodine is formed.

(9)   Titrate the iodine produced with the sodium thiosulfate solution, adding the thiosulfate until the colour of the solution in the conical flask becomes pale yellow. Then add approximately 1 cm³ of the starch indicator and complete the titration.

(10) Repeat the titration to obtain two consistent results.

Procedure total: 7 marks

■ ■ ■

Candidates' responses to A2 task 1

Candidate A

Mass of container + copper sulfate crystals/g	31.64
Mass of container/g	25.47
Mass of copper sulfate crystals/g	6.17

	Trial	1	2	3
Final burette reading/cm³	26.75	26.20	26.10	26.20
Initial burette reading/cm³	0.00	0.00	0.00	0.00
Titre/cm³	26.75	26.20	26.10	26.20
Titres used to calculate the mean/cm³		✓		✓
Mean titre value to 1 d.p./cm³	26.2			

Candidate B

Mass of container + copper sulfate crystals/g	32.12
Mass of container/g	25.87
Mass of copper sulfate crystals/g	6.35

	Trial	1	2	3
Final burette reading/cm³	27.00	26.65	26.40	26.30
Initial burette reading/cm³	0	0	0	0
Titre/cm³	27.00	26.65	26.40	26.30
Titres used to calculate the mean/cm³		✓	✓	✓
Mean titre value to 1 d.p./cm³	26.45			

Candidate A scores the mark for recording the masses but Candidate B makes a subtraction error. The mass of copper sulfate crystals is 6.25 g, not 6.35 g. Subtraction errors are surprisingly common. You should always check your working.

Both candidates score the mark for the accuracy of their titrations, although Candidate A has carried out three accurate titrations when the first two (26.20 cm³ and 26.10 cm³) were in good enough agreement for this not to be necessary. Candidate A scores the second mark for correct presentation of the titration results. Candidate B loses the mark for the initial titration reading by writing 0, rather than 0.00. Candidate A quotes the mean titre to one decimal place for 1 mark, but Candidate B loses this mark for two reasons. First, all the titration results (except the trial) have been used to obtain the mean instead of the two closest results. Three results should only be used if the three titres are spaced equally (e.g. 26.60 cm³, 26.40 cm³, 26.20 cm³). Second, the mean titre is not given to one decimal place. It may be that Candidate B thought that because 26.45 cm³ is midway between 26.4 cm³ and 26.5 cm³ the answer could not be given to one decimal place. However, the rule is to round up when this happens. So the answer is 26.5 cm³.

Both candidates carried out the titration carefully and score the 3 marks for obtaining a result that, after taking the different masses used into account, agrees with the teacher's mean titre. (The teacher would correct the subtraction error made by Candidate B and use the corrected figure to determine the result that should have been obtained.)

Overall, for carrying out the experiment, Candidate A scores the full 7 marks while Candidate B scores only 4 marks. This is despite having carried out the experiment well. All Candidate B's marks were lost through carelessness.

Analysis:

(a) The equations for the reactions taking place are:

$$2Cu^{2+}(aq) + 4I^-(aq) \rightarrow Cu_2I_2(s) + I_2(aq)$$
$$I_2(aq) + 2S_2O_3^{2-}(aq) \rightarrow S_4O_6^{2-}(aq) + 2I^-(aq)$$

Explain how these equations show that 1 mole of Cu^{2+} produces an amount of iodine that would react with 1 mole of $S_2O_3^{2-}(aq)$. (1 mark)

(b) Calculate the amount, in moles, of sodium thiosulfate contained in the mean titre. (1 mark)

(c) Calculate the amount, in moles, of copper sulfate in the solution you made up in the volumetric flask (2 marks)

(d) Use your result from (c) to determine a value of x. (2 marks)

(e) Use your result from (c) to calculate the percentage yield of the reaction between 5.00 g of copper oxide and sulfuric acid. (2 marks)

Analysis total: 8 marks

■ ■ ■

Candidates' responses

Candidate A

(a) 2 mol of Cu^{2+} produce 1 mol of iodine
1 mol of iodine reacts with $2S_2O_3^{2-}$
Therefore, 1 mol of Cu^{2+} will need 1 mol of $S_2O_3^{2-}$.

(b) mean volume = 26.20 cm^3

moles of $S_2O_3^{2-} = 0.1 \times \dfrac{26.20}{1000} = 0.00262$ mol

(c) 1 mol of Cu^{2+} needs 1 mol of $S_2O_3^{2-}$, so amount of Cu^{2+} is also 0.00262 mol
so, amount (in moles) in 250 cm^3 is 0.0262 mol

(d) 0.0262 mol weighs 6.17 g

1 mol weighs $\dfrac{6.17}{0.0262} = 235.5$ g

r.m.m. of $CuSO_4 = 63.5 + 32.1 + 64.0 = 159.6$ g

number of water of crystallisation $= \dfrac{(235.5 - 159.6)}{18.0} = 4.22$

(e) r.a.m. of Cu = 63.5 so 0.0262 mol weighs 1.664 g

% yield is $\left(\dfrac{1.664}{5.00}\right) \times 100 = 33.3\%$

Candidate B

(a) From the equations it can be seen that Cu^{2+} and $S_2O_3^{2-}$ have the same iodine.

(b) titre $= 26.45\,cm^3$

moles of $S_2O_3^{2-} = 0.1 \times \dfrac{26.45}{1000} = 0.0026\,mol$

(c) moles of Cu^{2+} are $0.026\,mol$

(d) 5

(e) mass of $CuSO_4 = 159.5\,g$

$0.026\,mol$ weighs $4.15\,g$

% yield $= \left(\dfrac{4.15}{5}\right) \times 100 = 83\%$

📝 Candidate A scores all the marks for parts (a), (b) (c) and (d) of the analysis. In part (e), the candidate becomes confused when calculating the percentage yield. Moles should be used, not grams. The calculation should be as follows:

amount of CuO used to prepare the crystals $= \dfrac{5.00}{79.5} = 0.0629\,mol$

100% yield would produce $0.0629\,mol$ of copper sulfate containing $0.0629\,mol$ of copper ions

$0.0262\,mol$ was obtained

so, the % yield is $\left(\dfrac{0.0262}{0.0629}\right) \times 100 = 41.7\%$

Candidate A gains 1 mark for part (e) because some progress was made.

Candidate B has more difficulty with the analysis. The answer to part (a) is close to being the correct explanation and it is possible that the student thinks this is an appropriate justification. However, it is not precise enough for a mark to be awarded. In part (b) the answer has been rounded to $0.0026\,mol$. This is an unnecessary and unacceptable approximation at this stage. The titration was more accurate than $26\,cm^3$. The candidate then uses the rounded answer for the next steps. This should *never* be done, Always maintain all the digits on your calculator for each step of the calculation, otherwise your final answer will almost certainly be incorrect. However, 2 marks are allowed for part (c) because the candidate uses the correct procedure. The steps in the calculation *should* be shown because even if an error is made some credit may still be given for the progress made. The answer to (d) is a guess. The candidate may have remembered that this is the correct value when the crystals are pure. Without any supporting working, it is impossible to deduce this, so a random guess never scores any marks. The answer to part (e) is worthy of just 1 mark for the correct molar mass of copper sulfate, but the rest of the calculation is wrong.

Overall, Candidate A loses only 1 mark and therefore achieves an excellent 14 marks out of 15. Candidate B only adds 3 more marks to the 4 scored for carrying out the experiment, giving a total of 7 marks. This is insufficient for a grade E.

A2 task 2: the pH of a buffer solution

It is possible that some tasks will require more than one practical session to complete. In this case, you will be given separate instructions for the two sessions.

First practical session
You are provided with a solution, **X**, containing a mixture of ethanoic acid and calcium ethanoate. You will determine the volume of hydrogen given off when solution **X** reacts with excess magnesium. You will then collect the calcium carbonate precipitated when solution **X** reacts with excess sodium carbonate solution.

All readings should be recorded in the tables given.

Part 1 procedure:
(1) Set up the apparatus ready to collect a gas over water in a $100\,cm^3$ measuring cylinder.
(2) Fill a burette with solution **X**.
(3) Use the burette to measure $15.0\,cm^3$ of solution **X** into a boiling tube.
(4) Add the magnesium powder provided and immediately connect the boiling tube to the apparatus to collect all the gas produced over water.
(5) Measure and record the volume of hydrogen obtained in the reaction.

Part 2 procedure:
(1) Use the burette to measure another $15.0\,cm^3$ of solution **X** into a clean boiling tube.
(2) Using a measuring cylinder, carefully add $10\,cm^3$ of the sodium carbonate solution provided. This is sufficiently concentrated to neutralise the ethanoic acid and precipitate all the calcium ions present as calcium carbonate.
(3) Weigh a piece of filter paper. Record the mass. Place the paper in a filter funnel.
(4) Filter the mixture in the boiling tube, taking care to transfer all the precipitate to the filter paper.
(5) Leave the filter paper and precipitate to dry.

Procedure total: 6 marks

■ ■ ■

Candidates' responses to A2 task 2
Candidate A

Final volume in the measuring cylinder/cm^3	86
Initial volume in the measuring cylinder/cm^3	0
Volume of gas collected/cm^3	86

Mass of filter paper = $1.56\,g$

Candidate B

Final volume in the measuring cylinder/cm³	75.2
Initial volume in the measuring cylinder/cm³	0.0
Volume of gas collected/cm³	75.2

Mass of filter paper = 1.53 g

Second practical session
Reweigh the dried filter paper and calcium carbonate. Now complete the results table.

Analysis:
You will determine the concentration of the ethanoic acid from the volume of hydrogen measured.

You will then determine the concentration of the calcium ethanoate from the mass of calcium carbonate precipitated.

Finally, you will determine the pH of solution **X**. The mixture is a buffer solution.

(a) **Write an equation for the reaction between magnesium and ethanoic acid.**

(1 mark)

(b) **Calculate the amount, in moles, of ethanoic acid in 15.0 cm³ of solution X.**
   (Assume that 1 mole of gas at r.t.p. has a volume of 24 dm³.) (1 mark)

(c) **Calculate the concentration of the ethanoic acid in solution X in mol dm⁻³.**

(1 mark)

(d) **Write an ionic equation for the precipitation of calcium carbonate from solution X.**

(1 mark)

(e) **Calculate the concentration of the calcium ethanoate in solution X.** (2 marks)

(f) **Use your answers to parts (c) and (e) to calculate the pH of the buffer solution X.**
   (K_a for ethanoic acid = 1.7 × 10⁻⁵ mol dm⁻³) (3 marks)

Analysis total: 9 marks

■ ■ ■

Candidates' responses

Candidate A

Mass of filter paper + calcium carbonate/g	2.23
Mass of filter paper/g	1.56
Mass of calcium carbonate/g	0.67

[?] Candidate A has completed the experiment reliably and recorded the results carefully. The results agree with those of the teacher, so the candidate scores all 6 procedure marks.

(a) $Mg + CH_3COOH \rightarrow CH_3COOMg + \frac{1}{2}H_2$

(b) moles of $H_2 = \dfrac{86}{24000} = 0.00358\,mol$

moles of $CH_3COOH = 2 \times 0.00358 = 0.00717\,mol$

(c) concentration of ethanoic acid $= 0.00717 \times \left(\dfrac{1000}{15.0}\right) = 0.478\,mol\,dm^{-3}$

(d) $Ca^+ + CO_3^- \rightarrow CaCO_3$

(e) moles of calcium carbonate $= \dfrac{0.67}{100.1} = 0.0067\,mol$

moles of calcium ethanoate $= 0.0067\,mol$

concentration of calcium ethanoate $= 0.0067 \times \left(\dfrac{1000}{15.0}\right) = 0.446\,mol\,dm^{-3}$

(f) $1.7 \times 10^{-5} = \dfrac{[H^+](0.446)}{(0.478)}$

$[H^+] = \dfrac{(8.13 \times 10^{-6})}{0.446} = 1.822 \times 10^{-5}$

$pH = 4.74$

Candidate B

Mass of filter paper + calcium carbonate/g	2.14
Mass of filter paper/g	1.53
Mass of calcium carbonate/g	0.61

[?] Candidate B obtained a satisfactory result for the mass of calcium carbonate but has been careless in collecting the hydrogen gas. The volume recorded is low, so some gas was probably allowed to escape before the measuring cylinder was in place. The candidate loses 1 mark. In addition, the volume in the measuring cylinder is recorded to an accuracy greater than that allowed by the equipment, so a further mark is lost. Therefore, Candidate B scores 4 procedure marks.

(a) $Mg + 2CH_3COOH \rightarrow (CH_3COO)_2Mg + H_2$

(b) moles of $H_2 = \dfrac{75.2}{24000} = 0.0031$

moles of CH_3COOH is twice as much $= 0.0062\,mol$

(c) concentration of ethanoic acid $= 0.0062 \times \left(\dfrac{1000}{15.0}\right) = 0.42\,mol\,dm^{-3}$

(d) $Ca^{2+}(aq) + Na_2CO_3(aq) \rightarrow CaCO_3(s) + 2Na^+(aq)$

(e) moles of calcium carbonate $= \dfrac{0.61}{100.1} = 0.0061\,\text{mol}$

formula of calcium ethanoate $= (CH_3COO)_2Ca$

so, moles of calcium ethanoate $= 2 \times 0.0061 = 0.0122\,\text{mol}$

concentration of calcium ethanoate $= 0.0122 \times \left(\dfrac{1000}{15.0}\right) = 0.813\,\text{mol}\,\text{dm}^{-3}$

(f) $1.7 \times 10^{-5} = \dfrac{[H^+](0.813)^2}{(0.42)}$

$[H^+] = \dfrac{(7.14 \times 10^{-6})}{0.813^2} = 1.08 \times 10^{-6}$

$pH = 5.97$

📝 Equations may be asked in all the task assessments but Candidate A still fails to give the correct formula for magnesium ethanoate. Candidate B does score the mark for part (a). Both candidates have the correct answer to part (b). Candidate A is lucky that the error in the formula of magnesium ethanoate does not affect the calculation. Three significant figures are really too many for the answer to part (b), which is limited by the accuracy of the measuring cylinder, but no mark would be lost. Both also obtain the mark for part (c). The ionic equation, required for part (d) causes difficulty for both candidates. Candidate A has incorrect charges on the ions and does not include state symbols, which are essential for an ionic equation. Candidate B does a little better but does not show the sodium carbonate solution in the form of ions. Neither candidate scores the mark.

The correct equation is as follows:

$$Ca^{2+}(aq) + CO_3^{2-}(aq) \rightarrow CaCO_3(s)$$

Candidate A does well, scoring both marks for part (e) but Candidate B is confused. Each mole of calcium ions from the calcium ethanoate reacts to produce 1 mol of calcium carbonate. The two ethanoate ions are irrelevant. However this is the only error, so Candidate B scores 1 mark.

Part (f) is difficult and neither candidate obtains the correct answer. Candidate A fails to realise that each mole of calcium ethanoate produces two moles of ethanoate ions. The concentration is, therefore, given by $2 \times 0.446 = 0.892\,\text{mol}\,\text{dm}^{-3}$. Otherwise, the method is correct and the candidate scores 2 marks. Candidate B uses an incorrect figure from part (e), which would be ignored if the rest of the calculation were correct. Unfortunately this is not the case, because Candidate B squares the ethanoate ion concentration, rather than doubling it. The hydrogen ion concentration is also out by a factor of 10, which is an easy mistake to make when working with small numbers. These two errors mean that only 1 mark is scored.

Overall, Candidate A has a respectable total of 12 marks out of 15 but Candidate B, through making careless mistakes, only achieves 9 marks.

Evaluative tasks

Nearly all students find that the evaluative task is the most demanding of the assessments. However, the questions set are based on procedures that you will have carried out during your practical sessions and, although you might not be able to prepare for the evaluative task in the same way as for the qualitative and quantitative tasks, it is still possible to equip yourself with the skills that are needed. As part of the evaluative task you will almost certainly be required to interpret the results of an experiment. Your teacher will tell you the topic on which the evaluation is based and you should make sure that you are familiar with the theory and any calculations that might form part of the assessment.

To succeed you must be able to:
- recognise weaknesses in the methods used in practical work. These are known as procedural errors.
- understand how the measurements that have been taken may limit the accuracy of any quantity that is calculated from the results. You may be asked to calculate the percentage error involved in a measurement taken using standard apparatus (see pp. 56–57).
- identify which errors, in either the procedure used or the measurements made, are the most significant
- anticipate how a change in the conditions used may affect any of the results that are obtained from an experiment

All this should be practised throughout the course. However, it is possible to review these points when looking back at experiments that you have already carried out.

Here, it is not possible to cover all the different types of question that might be asked within an evaluative task. Almost any aspect of the specification might be used as the basis for a question. However it is possible to focus on practicals that illustrate most of what may be required. These are listed below.

At both AS and A2 these are:
- generating and collecting gases
- using volumetric equipment — volumetric flasks, pipettes and burettes
- measuring the heat produced in a chemical reaction and deducing enthalpy changes
- the thermal decomposition of a solid
- measuring the rate of a reaction
- preparing a liquid organic compound using distillation
- interpreting mass spectra and infrared spectra as covered in Unit F322

At A2 only:
- preparing solid organic compounds
- interpreting proton and ^{13}C-NMR as covered in Unit F324
- understanding the use of chromatography
- providing more detail in the interpretation of rate experiments and, in particular, the calculation of orders of reaction

- understanding the behaviour of acids, bases and buffers, including the associated pH calculations
- understanding the measurement and use of electrode potentials
- understanding equilibria and equilibrium constants

At A2, many of the issues met at AS remain relevant and questions may require you to have an understanding of the weaknesses in procedures and measurements that were encountered then. The difference is that the experiments involved may need a more secure understanding of the background theory. Some general aspects of the application of the theory are reviewed in the relevant sections below. However, before attempting an evaluative task you should make sure that you have revised the topic thoroughly using either a textbook or your class notes.

Percentage errors

The estimation of percentage error in a quantity depends on whether it was measured by taking one or two readings. The volume measured using a pipette is a single reading, when the bottom of the meniscus of the solution is sitting on the mark on the pipette. The volume delivered by a burette is obtained by reading the burette twice — when you take the initial and final readings of a titration. There is usually no doubt as to whether one or two readings have been taken but, if there is any uncertainty, the wording of a question will always make it clear.

The expression 'maximum error' may be used in a task to indicate how accurate you can assume any measurement to be. For example, $50 \, cm^3$ of solution obtained using a measuring cylinder might be described as 'having a volume with a maximum error of $\pm 1 \, cm^3$'. This means that the actual volume can be said to lie between $49 \, cm^3$ and $51 \, cm^3$. The percentage error expresses the same idea but the error is quoted as a percentage of the volume taken — in this case, $\left(\frac{1}{50}\right) \times 100 = \pm 2\%$.

The percentage error in a quantity obtained by a *single* reading is:

$$\frac{\text{maximum error}}{\text{quantity measured}} \times 100$$

The percentage error in a quantity obtained from *two* readings is:

$$\frac{2 \times \text{maximum error}}{\text{quantity measured}} \times 100$$

The maximum error is sometimes, but not always, etched into the side of a piece of apparatus. As a general rule of thumb, for pieces of apparatus with a continuous scale (e.g. thermometers, measuring cylinders or burettes) you can take the maximum error to be half of a division of the scale used. If no other information is available this is a useful guide. When answering questions in the tasks it is essential to use the information provided. The following examples should make the calculation clear.

- **Pipette** — 25.0 cm³ measured with a maximum error of ±0.06 cm³ has a percentage error of $\left(\frac{0.06}{25.0}\right) \times 100 = \pm 0.24\%$. (As the maximum error is only quoted to one significant figure, this should, strictly, also be expressed to one significant figure, i.e. ±0.2%)

- **Burette** — 26.2 cm³ measured in a burette with a maximum error of ±0.05 cm³ in each reading has a percentage error of $\left(2 \times \frac{0.05}{26.2}\right) \times 100 = \pm 0.38\%$ (strictly, 0.4%).

- **Thermometer** — a temperature change measured from 19.5°C to 27.0°C using a thermometer with a maximum error of ±0.2°C in each reading has a percentage error in the temperature rise of $\left(2 \times \frac{0.2}{7.5}\right) \times 100 = \pm 5.3\%$ (±5%).

Generation and collection of gases

Two methods are commonly used to collect gases produced in a reaction (see Figures 3 and 4).

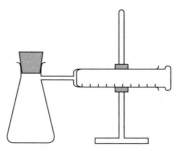

Figure 3 Collection of a gas using a syringe

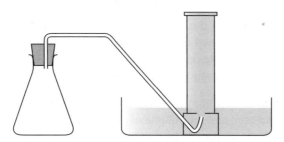

Figure 4 Collection of a gas over water

Either method is usually acceptable, although syringes are normally able to collect a maximum of 200 cm³ of gas while measuring cylinders are available in a variety of sizes up to above 1000 cm³. The diagrams show conical flasks being used, but if a solid is heated to generate a gas, a test tube or boiling tube should be used, rather than a conical flask.

If a gas is soluble in water and a measurement of its volume is required then a syringe must be used. Carbon dioxide is an example of a common gas that is sufficiently soluble in water that it is better to collect it in a syringe.

Using either method, the gas may escape before being collected. This is a particular problem if a solution has to be added to a substance in the flask in order to start the reaction — for example, when an acid is added to a carbonate to generate carbon dioxide or when an acid is added to a metal to release hydrogen. In these cases, the gas could be lost through the top of the flask before the bung has been replaced. The way to prevent this from happening is to place one of the reactants in a small tube inside the flask. The other reactant can then be added to the flask, the bung replaced, and the two components mixed by shaking the flask. This is illustrated in Figure 5 for the reaction between calcium carbonate and hydrochloric acid.

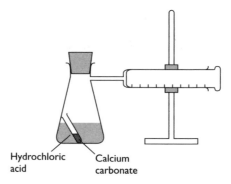

Hydrochloric acid   Calcium carbonate

Figure 6 Keeping two reactants separated in a flask

Sometimes students suggest placing the solid in the flask and running in the acid from a tap funnel. However, if this were done, the acid would displace air into the syringe as it was added. The volume measured would then include a volume of air equivalent to the sum of the volumes of the acid and the gas produced by the reaction.

It is worth remembering that the gas collected from an experiment of this kind will always contain some air pushed into the syringe by the gas that is generated in the reaction. This will not affect the result since the total volume of gas and air collected in the syringe will be identical to the volume of gas produced.

Errors resulting from the measurements made are those associated with the reactants — the mass of any solid used and the volume of any liquid or solution required — as well as the volume of the gas produced. Each of these has an uncertainty that is dependent on the accuracy of the equipment used. Some students make the assumption that a syringe is more accurate than a measuring cylinder. However, this is not necessarily the case and the information given in a question *must* be used.

An additional point to remember is that the volume of a gas depends on the temperature and pressure. An increase in temperature causes a gas to expand significantly. For example, $94\,cm^3$ of gas measured at 20°C expands to $120\,cm^3$ at a temperature of

100°C. If volumes of gases are to be compared it is, therefore, important to make the measurements at the same temperature. Many experiments involve the production of a gas as a result of heating a solid (e.g. the decomposition of a carbonate). The gas collected should be cooled to room temperature if a calculation is to be based on the assumption that one mole of gas has a volume of $24\,dm^3$.

The pressure in a laboratory does not vary much from day to day but even so it could make a difference to the measurement of a gas volume. The volume of a gas is inversely proportional to its pressure, so if a gas has a volume of $100\,cm^3$ at a pressure of $100\,kPa$, its volume would be just over $98\,cm^3$ if the pressure rose to $102\,kPa$. However, a change in the atmospheric pressure is rarely significant.

Volumetric experiments

When used correctly, volumetric flasks, burettes and pipettes are among the most accurate equipment in a laboratory. Their correct use is, therefore, almost certainly not going to contribute to significant error in an experiment. However, if they are used carelessly some of their precision is lost. You should be ready to answer questions that test your awareness of the consequences of misuse of apparatus. Examples include:

- A pipette must be filled so that, viewed directly, the bottom of the meniscus is precisely on the line etched on the apparatus. If the meniscus is under the line, the volume is less than intended.
- When a pipette is emptied, the surface of the solution that has been run out should be touched by the bottom of the pipette. If it is not, the volume measured is slightly less than the correct value. On the other hand, if the small volume of solution left in the bottom of a pipette is blown out, then the volume of solution obtained is greater than intended.
- A burette must be dry before it is filled, otherwise the solution poured into it will be diluted.
- If the funnel is not removed from the top of the burette while the titration is taking place solution may drip from it into the burette. This means that the volume of solution measured is less than it should be.
- If the solution from the burette is not mixed thoroughly into the solution in the conical flask, then extra solution from the burette may be needed to reach the end point.

You should understand that the conical flask does not have to be dried before the solution from the pipette is added. The presence of some drops of water will not alter the amount, in moles, added to the flask from the pipette. It is this amount that is being titrated and so the end point of the reaction will still occur when the correct volume of solution has been added from the burette. Of course, if it is an acid–base titration, anything in the conical flask must be neutral before the contents of the pipette are added.

Enthalpy experiments

Enthalpy experiments are among the least accurate experiments that you are likely to carry out. Difficulties in obtaining a reliable result tend to be obvious but nevertheless they are worthy of careful consideration.

Loss of heat to the surroundings is almost impossible to prevent and certainly occurs when a reaction takes place in a plastic cup. It is particularly significant when:

- a reaction is exothermic. Once the temperature rises above that of the surroundings, the reaction mixture loses heat readily and the temperature rise registered is less than it should be.
- a reaction occurs very slowly. Two solutions can normally be expected to react quickly together but if a solid is involved, the reaction is usually slower. Grinding the solid into a powder helps because the surface area of the solid is increased. Stirring the reaction mixture also helps, especially as the concentration of a solution decreases as the reaction progresses.

In both cases, better lagging of the reaction container and placing a lid on it help to reduce heat loss.

Remember that, if a reaction is endothermic and the temperature drops during the reaction, the problem is that heat may be absorbed from the surroundings, causing the fall in temperature to be less than expected.

If the temperature change is small, heat loss may be less significant. However, inaccuracy may then occur if the thermometer used is not sufficiently precise. A thermometer with a maximum error of $\pm 0.2°C$ used in measuring a temperature drop from $19.0°C$ to $17.4°C$ has a percentage error of $\left(2 \times \dfrac{0.2}{1.6}\right) \times 100$, which is $\pm 25\%$.

You may be asked what effect a change in concentration or conditions will have on the results of an experiment. This needs careful thought. Unless the reagents are in the exact reacting proportions as given by the equation, one or other of the reagents will be in excess. Increasing the concentration of this reagent will have little effect on the reaction. (The only change is that the reaction would go faster, so the heat lost to the surroundings might be different.) On the other hand, increasing the concentration of the reagent that is not in excess will increase the heat produced (until that reactant itself becomes in excess).

For example, the equation for the reaction between zinc and aqueous copper sulfate is:

$$Zn(s) + CuSO_4(aq) \rightarrow ZnSO_4(aq) + Cu(s)$$

When $0.010\,mol$ of zinc is added to $30\,cm^3$ of a solution containing $0.030\,mol$ of copper sulfate, the heat produced is x joules. Increasing the concentration of the copper sulfate still results in x joules being released as there is already enough copper sulfate solution to allow the reaction to go to completion. On the other hand, doubling the amount of zinc to $0.020\,mol$ would result in $2x$ joules being released.

Remember also that the heat is transferred to the total volume of the solution. If, in the reaction above, more moles of copper sulfate were supplied by doubling the volume of the copper sulfate to $60\,cm^3$, then the x joules would be spread through twice the volume of solution. Therefore, the temperature rise would be halved.

Thermal decomposition of solids

When carrying out a quantitative thermal decomposition experiment, great care is needed when heating the solid. In an evaluation task, you could be asked to assess what problems might arise. An experiment with which you should be familiar is the determination of the percentage of water of crystallisation in crystals, by heating them to expel all the water. The calculation is straightforward and, if the molar masses are known, it can be extended to determine the number of moles of water of crystallisation per mole of the anhydrous solid. To obtain accurate results requires awareness of the difficulties that might be encountered, and these will serve to illustrate the issues involved in this kind of experiment:

- The most obvious problem occurs when the crystals are not heated sufficiently for all the water to be removed. If this happens, the percentage of water of crystallisation calculated will be less than the correct value. The solution to this problem is to make sure that the crystals are heated to constant mass, i.e. a check should be made that further heating of the sample does not result in a further decrease in mass.
- There is also the possibility that the anhydrous solid may decompose further once the water has been removed (in some cases, before it has been completely removed). This a particular issue if the anion of the substance is a nitrate or a sulfate, as both break down fairly easily. It can also occur with chlorides because the water reacts with the chloride ion to produce hydrogen chloride gas. If this happens, the percentage of water of crystallisation measured will appear to be greater than the true value. To stop this happening requires careful temperature control during the heating. Water is released at 100°C, but most anhydrous solids require a higher temperature before they decompose.
- A further problem that may occur is the presence of impurities (left over from a previous experiment) on the surface of the vessel used to heat the crystals. These may decompose when heated and result in a loss in mass. However, unless you are told otherwise, you can assume in an evaluation task that all the apparatus is clean and dry. It is *never* the correct answer to a question to suggest that dirty apparatus is the cause of a problem.
- It is almost certain that a hot anhydrous solid will absorb some water as it cools. Yet it is not acceptable to weigh a hot solid on an accurate balance. If this problem is not solved, the extra weight resulting from the absorption of water will lead to an inaccurate result. The solution is to keep the solid in dry conditions as it cools, preferably in a desiccator (a piece of equipment designed for this purpose).

- Correct heating means avoiding the use of the yellow flame of a Bunsen burner. This causes carbon to be deposited on the surface of the container being heated, which will be added to the weight of the anhydrous product.

Measurement of the rate of reaction

There are only a few experiments to measure the rate of reaction that can be carried out readily as a quantitative task. (There are more at A2 — see p. 83.) However, you could be expected to show understanding of the issues involved in such experiments as part of an evaluative task. This requires little more than remembering the points covered in theory lessons.

The rate of a reaction depends on a number of factors, including:

- the **concentrations of the reactants**, if solutions are being used. You should be aware that the exact effect may not always be easy to predict and that occasionally changing a concentration has little or no effect on the rate.
- the **surface area** of any solid that is used. A powder reacts much faster than solid lumps of the same substance.
- the **temperature** of the reactants. There is a rule of thumb which states that the rate of a reaction doubles for every 10°C rise in the temperature. Although this is not always true, it does emphasise that temperature is an important factor.
- the **pressure**. Remember that changing the pressure only has a noticeable effect if a gas is involved.

Questions on this topic might require you to interpret the results of a change in the conditions or the amount of a reagent. You might be asked to sketch the effect of the change on a graph of concentration against time. There are two things to consider:

- Will the change increase or decrease the rate of the reaction? This is usually straightforward. An increase in temperature always increases the rate of a reaction; an increase in concentration usually increases the rate.
- Will the amount of product obtained be different? This is easy to forget but is often an important part of the answer required. If a reagent is in excess, increasing its concentration will not affect the amount of product obtained. If a reagent is not in excess, increasing its concentration will increase the amount of product until the reagent is no longer in excess.

Preparation of organic compounds

Liquid organic compounds

Although you might not have had the opportunity to carry out experiments involving the preparation and purification of an organic liquid, the procedures involved should be understood. The steps required are refluxing (not always necessary), initial distillation, purification and, finally, re-distillation within a specific temperature range. You are expected to understand the reasons for the various steps.

Refluxing is used when the reagents are slow to react and, therefore, need to be mixed continuously at a raised temperature. This is achieved by placing a condenser above a heated flask as illustrated in Figure 6. In the diagram, ethanol is being refluxed to ensure that all the alcohol is fully oxidised to ethanoic acid.

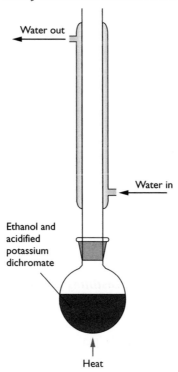

Figure 6 Refluxing a mixture

Notice that the water has to enter at the bottom of the condenser to make sure that it fills the outside jacket. If it were fed from the top of the condenser, the water would spill downwards without ever completely filling the jacket. During refluxing, the alcohol evaporates into the condenser where the vapour cools, becomes liquid again, and then returns to the flask below. Heating the flask must be controlled so that the vapour does not escape from the top of the condenser. It should never be allowed to go further than halfway up the condenser. A stopper must *not* be placed in the top of the condenser because the expanding gases would blow it out (or, if it was very tight, the apparatus might be put under excessive pressure, resulting in an explosion).

For the distillation steps, the condenser is set up as shown in Figure 7 on p. 64. The diagram shows the use of acidified dichromate to oxidise ethanol to ethanal. The distillation drives off the ethanal quickly and thus avoids most of the further oxidation to ethanoic acid.

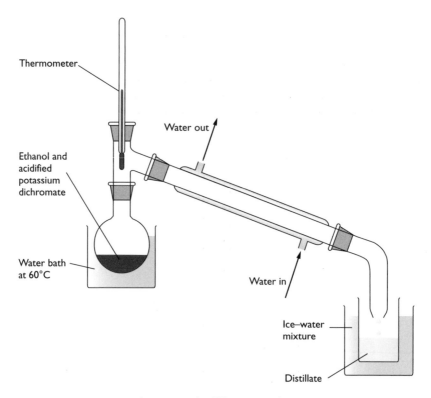

Thermometer

Water out

Ethanol and
acidified
potassium
dichromate

Water bath
at 60°C

Water in

Ice–water
mixture

Distillate

Figure 7 Distilling a product

For the first distillation of the product it is not necessary to insert the thermometer, but the thermometer is required for the re-distillation after the product has been purified. For this step, the thermometer should be placed so that its bulb takes the temperature of the vapour just before it enters the condenser.

The liquid produced by the first distillation will be impure; it is likely to contain some of the reactants used and will almost always contain water from the solutions used. For example, the distillate from the preparation of ethanal will contain some ethanol, water and acid as well as ethanoic acid produced by the further oxidation of the ethanal. The purification stages remove as many of the unwanted substances as possible. You are not expected to know all the ways that impurities might be removed but you should be familiar with a typical procedure, as explained below.

Many organic compounds are immiscible (i.e. do not mix) with water. Therefore, it may be possible to remove an impurity by using an aqueous solution that reacts with the impurity. The impurity dissolves in the water, forming a separate liquid layer from the distillate. For example, if the unwanted compound is acidic, shaking with sodium carbonate solution causes the acid to react producing carbon dioxide and the sodium salt of the acid. The sodium salt dissolves in the water and two liquid layers are produced. Separating the aqueous solution from the organic product is achieved by using a separating funnel to run off the unwanted aqueous layer, as shown in Figure 8.

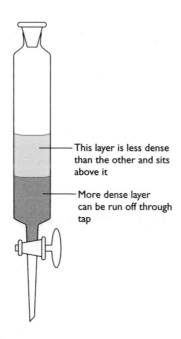

Figure 8 Using a separating funnel

Once the impurities have been removed, the distillate must be dried. This can be carried out using a number of drying agents, for example solid anhydrous sodium sulfate or magnesium sulfate. The drying agent is then filtered off and the filtrate containing the desired product is re-distilled using a thermometer to monitor the temperature of the vapour entering the condenser. The vapour produced around the boiling point of the liquid being prepared is then condensed and collected.

Preparation of solid organic compounds

This is required for A2 only.

The preparation of a solid organic compound might be included as part of a qualitative task, but an evaluative task might test an understanding of the reasons for some of the practical procedures. These could be specific to the preparation of a particular substance, but there are some general issues. To obtain a pure product, recrystallisation is usually required. A solvent is chosen in which the desired product dissolves readily at higher temperatures but only slightly at room temperature. The solvent is often water but it could be another suitable liquid, such as an alcohol. The impure product is dissolved in the hot solvent and is then separated by filtration from the insoluble impurities. On cooling, the product forms as crystals and is separated from any impurities that are sufficiently soluble to stay dissolved in the solvent.

The purity of the crystals obtained could be checked using spectroscopy, but a measurement of the melting point is also a sensitive check on purity. A pure substance melts over a narrow range of temperatures; an impure substance starts melting well below the true melting point and then continues to melt over a wide temperature range.

Spectroscopy and chromatography

Mass spectra and infrared spectra

Understanding how to interpret spectra should be straightforward and an evaluation task will test your use of the information you are given. This might include using a mass spectrum to establish the relative molecular mass of a compound and recognising some of the fragments obtained. For an infrared spectrum, you should be able to use the table in your data sheet to recognise the peaks for a C–O bond, a C=O bond, an –OH group in alcohols and an –OH group in carboxylic acids.

NMR spectroscopy and chromatography

This is required for A2 only.

Both ^{1}H-NMR and ^{13}C-NMR provide useful information that enables structural details of organic compounds to be established and, in some cases, allows a positive identification to be made. Nothing further than the ideas covered in Unit F324 is expected in an evaluative task. Before attempting any task that is based on organic chemistry, you should check that you understand the use of these spectra.

The principles of thin-layer and gas chromatography should also be understood. You should appreciate that both techniques allow the components of different mixtures to be separated. The effectiveness of chromatography is dependent on finding a stationary phase that allows good separation to occur. For example, in gas chromatography the retention times of compounds may be so similar that separation is ineffective. However, no further understanding of this topic is likely to be required.

Determination of the orders of reaction

This is required for A2 only.

An evaluation is likely to focus on whether the results obtained from a rate-determination experiment are accurate enough to allow valid conclusions to be made. It can be difficult to make reliable measurements in experiments of this type — assumptions often have to be made and these need to be understood.

Many rate investigations use a procedure in which the starting concentrations of the reactants are changed systematically to determine the effect that this has on the initial rate. As has been mentioned on p. 29, this is usually measured by taking the rate to be $1/t$ where t is the time taken to reach a particular point in the reaction. This might, for example, be the formation of a certain depth of colour or extent of a precipitation.

In fact, the rate is only proportional to $1/t$ for reactions that proceed at a constant rate over the time measured. This is never completely true because reactions always start faster and then slow down as the concentration of the reagents diminishes.

It is a good approximation if the time measured is short because the reaction rate will not have changed much. It is unreliable if the reaction is allowed to proceed too far before the chosen time, t. It is fortunate that most reactions (and all those you meet at A-level) are zero, first or second order with respect to the concentrations of the reagents. Even though an approximation is used in the measurement of the rate, the points on a graph of rate against concentration should still be accurate enough to allow the order with respect to a reagent to be identified correctly.

An alternative method of determining the order of a reaction is to allow it to run through its course and then to analyse the shape of a graph of concentration versus time for that particular reactant. However, reactions usually involve more than one reactant, all of which affect the rate. So to determine the order with respect to any one reactant a 'trick' is used to make sure that the other components do not have much effect on the overall rate. This 'trick' is to make sure that the reactant being investigated is present at a much lower concentration than any of the others. That way, the concentrations of the other reactants remain almost constant throughout the course of the reaction. Therefore, the rate of reaction can be assumed to be due to the reactant in low concentration. Suppose that the order with respect to X is being determined in a reaction involving three solutions, X, Y and Z. The rate equation is

rate $= k[X]^x[Y]^y[Z]^z$

If the concentrations of Y and Z are so large compared with X that their concentrations can be taken to be constant throughout, then the rate equation can be simplified to

rate $= k_1[X]^x$

and the order with respect to X can be determined.

Acids, bases and buffers

This is required for A2 only.

You should know the principles of acid–base reactions in terms of the Brønsted–Lowry theory and you should be able to identify and provide equations for acid–base reactions. Calculations of the pH of strong acids, strong bases, weak acids and buffer solutions may be asked within the context of an experiment. You should be aware of the approximations used in these calculations and how they could affect the results. These include:

- The ionic product of water is only exactly $1 \times 10^{-14}\,mol^2\,dm^{-6}$ at a particular temperature and increases as the temperature is raised.
- The pH of a weak acid is usually calculated using the approximation that the equilibrium concentration of the molecular form of the acid is the same as the concentration of the acid that was dissolved. This is not strictly true because some of the acid ionises. If the acid is not particularly weak the approximation is less good and it leads to the calculated pH value being higher than is correct.

- The pH of a buffer solution is usually calculated using two approximations. The first is that, at equilibrium, the anion has the same concentration as the anion of the salt before equilibrium. The second is that, at equilibrium, the acid has the same concentration as that of the acid dissolved initially. In fact, the acid is slightly ionised, which reduces the concentration of the acid. However, by supplying some extra anions as a result of the ionisation, the concentration of the salt is increased. The approximations are quite good because a further factor is that the anions from the salt suppress the ionisation of the acid. Nevertheless, to a small extent, it will lead to the calculated pH being higher than is correct.

You should be aware of how the shape of a pH curve during a titration identifies the strength of the acid and/or base. The choice of indicator for these titrations is important and you should understand that a reliable result will not be obtained if the indicator range is not suitable for the titration.

Electrode potentials and redox equations

This is required for A2 only.

Electrode potentials allow you to determine the feasibility of redox reactions. For evaluation tasks, you should understand how electrode potentials can be used (see any standard textbook) and also appreciate their limitations. Remember that standard electrode potentials are measured under specific conditions:

- a temperature of $298\,K$ ($25°C$) and a pressure of $101\,kPa$
- concentrations of $1\,mol\,dm^{-3}$ for all the ions involved in the cell

In most circumstances, a temperature of $298\,K$ is higher than normal laboratory conditions and the concentrations used are often less than $1\,mol\,dm^{-3}$. This is not as severe a limitation as it might seem. Electrode potentials vary only slightly with both temperature and concentration. For example, the standard electrode potential for $Cu^{2+}(aq) + 2e^- \rightarrow Cu(s)$ is $+0.34\,V$. Reducing the concentration of copper ions to $0.1\,mol\,dm^{-3}$ only reduces the electrode potential to $+0.31\,V$. With the concentration as low as $0.001\,mol\,dm^{-3}$, the value is $+0.25\,V$. Other electrode potentials may change by a little less or a little more from their standard values. However, unless there is a very large change in the concentration of the ion, these changes will be quite small. Therefore, predictions based on the use of electrode potentials are usually valid. It is, however, important to understand that to say a reaction is feasible does not mean it will necessarily occur quickly. The rate depends on the activation energy of the reaction and this is not related to the electrode potential. A reaction that is shown by the electrode potential values to be comfortably feasible may occur slowly, while another reaction which appears, on the basis of electrode potentials, to be only just feasible may take place quite rapidly.

If a task is based on electrode potentials you might be asked to write an overall ionic equation for the reaction by combining the two half-equations for a process. This essential skill should be practised thoroughly.

Equilibria and equilibrium constants

This is required for A2 only.

The use of Le Chatelier's principle to predict the effect of a change in concentration, temperature or pressure on the components present in an equilibrium mixture should be understood. It is unlikely that you would be asked to determine an equilibrium constant as a quantitative task because many reactions take a long time to establish a true equilibrium. However, you could be asked to interpret an experiment based on information provided. You should be prepared to do this and to understand how the procedure might be affected by a change in conditions. Remember, only a change in temperature alters the value of an equilibrium constant, but how this is altered depends on whether the reaction is exothermic or endothermic. The necessary background can be found in any standard textbook.

Exemplar evaluative tasks

The evaluative tasks vary, so it is difficult to give typical examples. Nonetheless, the following tasks illustrate the level of understanding that is expected and include many of the errors that are seen. The task is marked out of 15 but in some of these examples the mark is lower to avoid unnecessary duplication of similar types of question.

AS task 1: enthalpy of decomposition of copper carbonate

A student decides to determine the enthalpy change that occurs when copper carbonate decomposes into copper oxide and carbon dioxide.

This enthalpy change cannot be readily measured directly in the laboratory. However, since copper carbonate and copper oxide both react with sulfuric acid, the enthalpy change when each of these reactions takes place can be determined by simple experiments.

The equations for the two reactions are:

$$CuCO_3(s) + H_2SO_4(aq) \rightarrow CuSO_4(aq) + CO_2(g) + H_2O(l)$$
$$CuO(s) + H_2SO_4(aq) \rightarrow CuSO_4(aq) + H_2O(l)$$

Hess's law can be used to deduce the enthalpy change for the decomposition:
$$CuCO_3(s) \rightarrow CuO(s) + CO_2(g)$$

(a) Explain why it would be difficult to measure the enthalpy of the decomposition of copper carbonate directly in the laboratory. (1 mark)

(b) Write an ionic equation for the reaction between copper oxide and sulfuric acid. Include state symbols in your answer. (2 marks)

The student plans to carry out the reactions with copper carbonate and copper oxide and suggests the following procedure:

- In both experiments a measuring cylinder will be used to measure out $100\,cm^3$ of $2.00\,mol\,dm^{-3}$ sulfuric acid. This will be poured into a plastic cup and the temperature of the acid will be measured.

- For the first experiment, $2.47\,g$ of copper carbonate will be added to the sulfuric acid and the maximum temperature reached when the solid reacts will be measured. For the second experiment, this procedure will be repeated using $2.47\,g$ of copper oxide instead of copper carbonate.

(c) (i) The relative formula mass of $CuCO_3$ is 123.5.

Show, using a calculation, that in the first experiment the amount, in moles, of copper carbonate and the amount, in moles, of sulfuric acid that the student intends to use are the same. (2 marks)

(ii)   Explain why it would be better to carry out the experiment using a greater volume of sulfuric acid.   (2 marks)

(iii) The relative formula mass of copper oxide is 79.5. Explain why this means that in the second experiment the copper oxide would be in excess.   (2 marks)

(iv)   The reaction of copper oxide with sulfuric acid takes place very slowly at room temperature. Explain why the measurement of the temperature rise obtained from the reaction would be unreliable.   (1 mark)

The student discovers that there is no copper oxide available so decides to make some by heating copper carbonate until it is decomposed completely.

(d) How could the student tell when all the copper carbonate had decomposed?
                                                                                        (1 mark)

The student then carries out the two experiments to measure the enthalpy changes. In each case, the solid used for the experiments is reduced to 1.00 g. The sulfuric acid is now in excess for both experiments.

For the reaction of copper carbonate and sulfuric acid, the temperature rise is from 20.0°C to 25.5°C. Both thermometer readings have a maximum error of ±0.5°C.

(e) Calculate the percentage error in the measurement of the temperature rise.
                                                                                        (2 marks)

The student completes both experiments and uses the results to determine a value for the enthalpy of decomposition of copper carbonate. However, the value obtained by the student is less than the true value because the procedure used for the experiment is unsatisfactory.

(f) Apart from using more accurate measuring equipment, suggest two different ways in which the student might improve the experiment.   (2 marks)

Total: 15 marks

■ ■ ■

Candidates' responses to AS task 1

Candidate A

(a) The activation energy for the reaction is too high.

Candidate B

(a) Because the copper carbonate would need to be heated.

✔ Both candidates have written just enough to gain the mark. However, both should have explained that it would be almost impossible to determine a temperature change as a result of the reaction because heat has to be applied to cause the decomposition.

Candidate A

(b) $O^{2-}(s) + 2H^+(aq) \rightarrow H_2O(l)$

Candidate B

(b) $O^-(s) + H_2^+(aq) \rightarrow H_2O(l)$

Candidate A has written an ionic equation which is sufficient for 2 marks. However, the answer is not completely correct. Copper oxide is a solid lattice and the oxide ions are not separated from the copper ions. The correct ionic equation is:

$$CuO(s) + 2H^+(aq) \rightarrow H_2O(l) + Cu^{2+}(aq)$$

Candidate B fails to score because the charge on the oxide ion is incorrect and writing H_2^+ implies that the two hydrogen atoms are bonded in some way and that together they have a charge of $1+$.

Candidate A

(c) (i) They are both 0.02 mol.
(ii) If the sulfuric acid is not in excess it will be used up and so towards the end the reaction will be very slow and the temperature rise will not be recorded accurately.
(iii) The CuO has a lower r.m.m. than the copper carbonate. So 2.47 g is more moles and therefore must be more than the moles of sulfuric acid.
(iv) A lot of heat will be lost to the surroundings and the plastic cup.

Candidate B

(c) (i) moles of copper carbonate $= \dfrac{2.47}{123.5} = 0.02\,\text{mol}$
moles of acid $= \left(\dfrac{1}{10}\right) \times 2 = 0.02\,\text{mol}$
(ii) So that all the copper carbonate is used up in the reaction.
(iii) The molecule of copper oxide is smaller than that of copper carbonate, so it forms more moles. So, there will be more of it than the sulfuric acid.
(iv) Because the reaction would take a long time.

Both candidates obtain the 2 marks for the calculation in (c)(i) but Candidate A should have shown how the calculation was carried out. Candidate A's answer to part (c)(ii) gains 1 mark for explaining that the reaction slows down. However, the second point — that this would cause more heat loss and therefore an inaccurate temperature rise — is not explained adequately. It is important to give full details when providing explanations. Candidate B might be implying the correct answer to (c)(ii) but what is written would not gain any marks. The correct reaction quantities are present so the reaction would eventually go to completion. Candidate A's answer to (c)(iii) is sufficient for both marks but, strictly, r.m.m. should not be used. First, an ionic compound does not contain molecules and second, it is unwise to use abbreviations in answers. Relative formula mass is the correct term. Candidate B has confused mass and size and fails to score. Candidate A has answered question (c)(iv) satisfactorily, for 1 mark. Candidate B repeats the question and fails to score. The point required is that, if a reaction is slow, a lot of heat is lost to the surroundings and so the temperature rise is less than the true value.

Candidate A

(d) Heat the copper carbonate until its mass doesn't change any more.

Candidate B

(d) Copper carbonate is green and copper oxide is black. So, heat it until all the colour has changed to black.

Candidate A gives a good answer and scores the mark. Candidate B has remembered that there is a colour change in the reaction. However, this is not a sufficiently reliable way of deciding when the reaction is complete because the black colour could disguise the presence of some green powder that had not decomposed.

Candidate A

(e) Each reading has an error of ±0.5°C. Percentage error is $\left(\frac{1}{5.5}\right) \times 100 = 18.2\%$.

Candidate B

(e) error in 20°C = $\left(\frac{0.5}{20}\right) \times 100 = 2.5\%$

error in 25.5°C = $\left(\frac{0.5}{25.5}\right) \times 100 = 1.96\%$

total error = 2.5% + 1.96% = 4.46%

Candidate A does the calculation correctly and gains the 2 marks. However, showing more working would have been wise. The student realises that the error of ±0.5°C applies to both the 20°C and 25.5°C readings and therefore the temperature change of 5.5°C has a total error of ±1.0°C. Candidate B does not understand that the error relates to the temperature change that is measured and not to the individual readings. The candidate's answer is incorrect and scores no marks.

Candidate A

(f) Surround the cup with insulation and put a lid on the top with a hole for the thermometer.
Stir the solution with the thermometer.

Candidate B

(f) Surround the cup with cotton wool.
Grind the powders up.

Candidate A gives two improvements, for 2 marks. Candidate B scores 1 mark for the first improvement, but grinding substances that are already powdered would make little difference.

Although Candidate A has not answered the questions in the best way, an excellent 14 out of 15 marks is achieved. Candidate B seems to understand some of the work, but scores only 4 marks.

AS task 2: temperature change in the reaction between calcium carbonate and hydrochloric acid

When 1.00 g of calcium carbonate is added to 50.0 cm³ of 0.800 mol dm⁻³ hydrochloric acid in a plastic beaker the temperature of the acid rises from 19.5°C to 24.3°C.

(a) (i)   Write an equation for the reaction. Include state symbols.   (2 marks)

   (ii) Calculate the amount, in moles, of calcium carbonate used in the experiment.
                                                                          (1 mark)

   (iii) To show that it is present in excess, calculate the amount, in moles, of hydrochloric acid used in the experiment.   (1 mark)

(b) Explain what temperature rise you might expect to see if the experiment was repeated using:

   (i)   2.00 g of calcium carbonate and 50.0 cm³ of 0.800 mol dm⁻³ hydrochloric acid.   (2 marks)

   (ii) 4.00 g of calcium carbonate and 50.0 cm³ of 0.800 mol dm⁻³ hydrochloric acid.   (2 marks)

Total: 8 marks

■ ■ ■

Candidates' responses to AS task 2

Candidate A

(a) (i) $CaCO_3(s) + 2HCl(aq) \rightarrow CaCl_2(aq) + CO_2(g) + H_2O(l)$

   (ii) moles of $CaCO_3$ = 40 + 12 + 48 = 100 g

         $1.00 g = \frac{1}{100} = 0.01 mol$

   (iii) moles of HCl = $\left(\frac{50}{1000}\right) \times 0.8 = 0.04 mol$

Candidate B

(a) (i) $CaCO_3 + HCl \rightarrow CaCl + CO_2 + H_2O$

   (ii) 0.01 mol

   (iii) 0.04 mol

Candidate A completes part (a) well. The only issue is that the student has not used atomic masses from the periodic table provided. These are to one decimal place and the r.a.m. of calcium is 40.1 not 40. It should be noted that in some cases a mark would be lost for using a rounded atomic mass. However, as there is only 1 mark for this part-question, it would probably be awarded. Candidate A scores all 4 marks. Candidate B's equation is incorrectly balanced and the state symbols are not included, so the candidate fails to score. Although no working is shown for parts (ii) and (iii), the 2 marks are obtained for the correct answer. However, it is always sensible to show your working. Neither candidate is quite correct in the use of significant figures in parts (ii) and (iii). The information supplied was to three significant figures and both give answers to just one significant figure (e.g. 0.04, rather than 0.0400). The question will usually warn you if significant figures are required. Nevertheless, it is good practice to be consistent.

Candidate A

(b) (i) The heat produced is caused by the calcium carbonate reacting. Therefore doubling the mass of calcium carbonate will produce double the heat. The temperature rise will be twice as much — $2 \times 4.8 = 9.6°C$.

(ii) Only 2 g of the calcium carbonate can react before the acid runs out. So the temperature rise will once more be 9.6°C.

Candidate B

(b) (i) Twice as much calcium carbonate means double the enthalpy, so the temperature rise is 10.4°C.

(ii) Four times the calcium carbonate gives four times the enthalpy = 20.8°C.

Candidate A's answers to parts (b)(i) and (ii) gain all 4 marks. Candidate B has subtracted the thermometer readings incorrectly and obtained an incorrect temperature rise of 5.2°C. Therefore, the candidate might understand that the temperature rise has doubled but this is far from clear and the examiner would not give this the benefit of the doubt and award a mark. It is essential to avoid careless errors of this type. The second mark cannot be awarded because the candidate uses the word 'enthalpy' incorrectly. The enthalpy of a reaction is a fixed quantity and does not alter. It is the heat produced that depends on the quantities used. Part (ii) is not understood and no marks are obtained. (The incorrectly balanced equation suggests that there would be enough acid to achieve a temperature rise of four times the original. Had the candidate referred to the equation in the answer then 1 mark might have been allowed.)

Overall, Candidate A does really well and achieves the full 8 marks. Candidate B scores a disappointing 2 marks. The candidate needs to think questions through much more carefully.

AS task 3: determination of the relative molecular mass of a vapour

Two students want to determine the relative molecular masses of the two organic liquids shown in the table below.

Substance	Boiling point/°C
Ethanol	78.6
Octane	125.8

The method they plan to use for each liquid is:

- Weigh a sample of liquid in a boiling tube.
- Place the boiling tube in a hot water bath.
- Heat the water bath until all the liquid in the boiling tubes has evaporated.
- Measure the volume of the vapour produced in a graduated syringe.

From the volume of the vapour measured they intend to determine the amount in moles contained in the mass weighed out and to use this to determine the relative molecular mass.

The apparatus they intend to use is shown below.

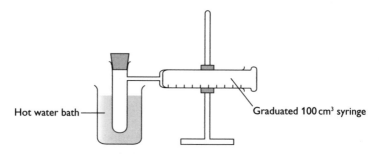

Hot water bath — Graduated 100 cm³ syringe

(a) Explain why the method they plan to use would make it impossible to determine the relative molecular mass of octane. (1 mark)

One of the students suggests that 1.00 g of ethanol should be used.

The relative molecular mass of ethanol is 46.0 and the volume of 1 mole of ethanol vapour at its boiling point is 31.0 dm³.

(b) Explain, by using a calculation, why 1.00 g is too large a mass to use with the apparatus illustrated. (3 marks)

The students decide to use 0.10 g of ethanol. The first student weighs the ethanol into the boiling tube and sets the syringe plunger to zero. He heats the water bath until all the ethanol has evaporated and the plunger of the syringe moves out. He decides to measure the volume in the syringe at room temperature and leaves the whole apparatus to cool. However, he is disappointed to find that the plunger returns to zero.

(c) (i) Explain why the plunger returns to zero. (2 marks)

The second student decides that it would be better to read the syringe as soon as all the ethanol has evaporated. The student weighs the ethanol into the boiling tube and sets the syringe plunger to zero. When all the ethanol has evaporated the student measures the volume in the syringe but finds that the volume collected is much greater than it should be.

(ii) Explain why the volume collected is greater than the true volume of 0.10 g of ethanol vapour at its boiling point. (2 marks)

The second student decides to check that all the ethanol vapour went into the syringe during the experiment and lets the syringe cool, expecting to see droplets of ethanol forming. However, there does not appear to be any ethanol in the syringe.

(d) Explain why it is unlikely that there will be much ethanol in the syringe. (2 marks)

Total: 10 marks

■ ■ ■

Candidates' responses to AS task 3

Candidate A

(a) The water can't get hot enough to evaporate the octane.

Candidate B

(a) Octane is petrol and would burn.

☑ Candidate A understands the difficulty, for 1 mark. Candidate B's comment is partly true but does not mean that the experiment is impossible.

Candidate A

(b) The number of moles of ethanol is $\frac{1}{46.0}$ = 0.0217 mol.

This has a volume of 0.0217 × 31.0 = 0.674 dm³ or 674 cm³.
The syringe can only hold 100 cm³, so much less has to be used.

Candidate B

(b) Moles of ethanol is $\frac{1}{46}$ = 0.02 mol

This has a volume of 0.02 × 24 = 0.48 dm³, which is too large a volume.

☑ Candidate A scores all the marks for a clearly expressed answer. Candidate B obtains 1 mark for following the correct procedure in the calculation. However, the candidate has not read the question carefully and uses 24 dm³, which is the volume of 1 mole of a gas at r.t.p. Many candidates make unnecessary mistakes through failing to look at the information provided. The candidate also rounds the amount, in moles, of ethanol to 0.02 mol. This is quite unjustified and leads to a large error in the calculation that follows. Finally, writing 'which is too large a volume' is not sufficiently clear to count as an adequate explanation as to why the mass is inappropriate. Reference to the syringe is required.

Candidate A

(c) (i) When the ethanol evaporates it becomes a gas and pushes the syringe out. When it cools, it condenses and the plunger goes back to its original position.

(ii) When the apparatus is heated the air inside it expands and the plunger will move out even before the ethanol has evaporated. This means that the gas collected will not just be the volume of ethanol vapour, it will also contain some air.

Candidate B

(c) (i) When the apparatus cools down it is easy for gas to escape and so the plunger will eventually leak enough to return to zero.

  (ii) As the ethanol evaporates it expands and pushes air with it. The volume in the syringe is the volume of ethanol plus the volume of air. The syringe might be at a greater temperature than the boiling point of ethanol.

In part (c)(i), Candidate A obtains 1 mark for saying that the ethanol condenses but fails to mention that the plunger only returns to zero when the air inside the apparatus has contracted fully. Candidate B tries to suggest a possible answer, but this will not gain marks. Always try to think of a positive reason why something occurs, rather than suggesting a failing in the apparatus as this is unlikely to be the response required. In part (c)(ii), Candidate A understands what is happening and gains the 2 marks. Candidate B might be given the benefit of the doubt for the first part of the answer to (c)(ii), although it is not clear that the reason has been understood. However, the second suggestion that the temperature in the syringe might be higher than the boiling point could not be true for the apparatus shown.

Candidate A

(d) The ethanol would readily evaporate from the syringe as it cools and therefore it is unlikely that droplets would be seen.

Candidate B

(d) 0.10 g of ethanol is a tiny amount, so it would be hard to see droplets forming.

Neither candidate answers this part well. Candidate B does not give the expected answer but as it is a reasonable suggestion it gains 1 mark. The expected answer is that the syringe would not fill with ethanol vapour during the experiment. It would contain air that had been displaced by the ethanol vapour.

Overall, Candidate A scores a creditable 7 marks out of 10. Candidate B only manages to score 3 marks.

A2 task 1: estimation of a solution containing $Fe^{3+}(aq)$ ions

Some students plan to use a titration to confirm that a solution they have been given has a concentration of $0.100\,mol\,dm^{-3}$ $Fe^{3+}(aq)$ ions.

Using a pipette, each student takes $25.0\,cm^3$ of the solution and places it in a conical flask. Each then carries out the procedure described below.

Students A and B add excess aqueous potassium iodide to the conical flask. Iodine is formed. They then titrate the iodine with aqueous sodium thiosulfate. This allows them to determine the amount of iodine present and hence the concentration of the $Fe^{3+}(aq)$ ions.

(a) (i) Use the electrode potentials below to show that the reaction of $Fe^{3+}(aq)$ ions with $I^-(aq)$ ions is feasible under standard conditions.

$$I_2(aq) + 2e^- \rightleftharpoons 2I^-(aq) \qquad +0.54\,V$$

$$Fe^{3+}(aq) + e^- \rightleftharpoons Fe^{2+}(aq) \qquad +0.77\,V \hspace{2cm} \text{(1 mark)}$$

   (ii) Write an ionic equation for this reaction.   (1 mark)

The equation for the reaction between iodine and sodium thiosulfate is shown below:

$$2S_2O_3^{2-}(aq) + I_2(s) \rightarrow S_4O_6^{2-}(aq) + 2I^-(aq)$$

Both students consider what would be the best choice of concentration for the sodium thiosulfate. Student A decides to use $0.100\,mol\,dm^{-3}$ sodium thiosulfate and Student B decides to use $0.0500\,mol\,dm^{-3}$ sodium thiosulfate.

(b) Use the equations to explain which student has made the better choice of concentration. Give clear reasons for your answer.   (3 marks)

Another student decides to reduce the $Fe^{3+}(aq)$ ions to $Fe^{2+}(aq)$ by adding a solution containing excess $Sn^{2+}(aq)$. The student then plans to titrate the $Fe^{2+}(aq)$ ions with acidified potassium manganate(VII) solution and determine the end point.

(c) Consider the following electrode potentials and then suggest why this method would not be suitable. (Solutions containing $Sn^{2+}(aq)$ and $Sn^{4+}(aq)$ ions are both colourless.)

$$Sn^{4+}(aq) + e^- \rightleftharpoons Sn^{2+}(aq) \hspace{4cm} +0.15\,V$$

$$Fe^{3+}(aq) + e^- \rightleftharpoons Fe^{2+}(aq) \hspace{4cm} +0.77\,V$$

$$MnO_4^-(aq) + 8H^+(aq) + 5e^- \rightleftharpoons Mn^{2+}(aq) + 4H_2O(l) \qquad +1.51\,V \quad \text{(2 marks)}$$

Total: 7 marks

Candidates' responses to A2 task 1

Candidate A

(a) (i) $Fe^{3+}(aq) + e^- \rightleftharpoons Fe^{2+}(aq)$ is $+0.77V$ which is more than the $-0.54V$ required for the reaction $2I^-(aq) \rightleftharpoons I_2(aq) + 2e^-$ so the reaction is possible.

(ii) The overall equation is $2Fe^{3+}(aq) + 2I^-(aq) \rightarrow 2Fe^{2+}(aq) + I_2(aq)$

Candidate B

(a) (i) The overall voltage for these reactions is $0.77 + 0.54 = 1.31$ V. This means the reaction will work.

(ii) The equation is $2Fe^{3+} + 2I^- \rightarrow 2Fe^{2+} + I_2$

✍ Candidate A's answer to part (a)(i) is correct, for 1 mark. Candidate B has just added the two electrode potentials together without considering which reaction supplies electrons and which receives them. The candidate fails to score. In part (ii), both candidates have the correct ionic equation but candidate B has not included state symbols. However, as these were not specifically asked for, the mark is awarded.

Candidate A

(b) From the equation above, $2Fe^{3+}$ produce 1 I_2.

1 I_2 reacts with $2S_2O_3^{2-}$ so every mole of Fe^{3+} you start with requires 1 mole of $S_2O_3^{2-}$ in the titration.

If the Fe^{3+} and the $S_2O_3^{2-}$ are the same concentration of $0.100\,mol\,dm^{-3}$ then $25\,cm^3$ of each will be required and the titration will be balanced.

Candidate B

(b) 2 moles of Fe^{3+} produce 1 mole of iodine, so the thiosulfate should be half the concentration of the Fe^{3+}. Therefore, $0.0500\,mol\,dm^{-3}$ is best.

✍ Candidate A scores all 3 marks for a well-considered answer. Candidate B however has not considered the reacting proportions of the iodine and thiosulfate. Candidate B fails to score.

Candidate A

(c) The Sn^{2+} would react with the MnO_4^-.

Candidate B

(c) Sn^{2+} is able to reduce the Fe^{3+} so the method ought to work. Perhaps the reaction is too slow to take place.

✍ It is difficult to tell whether Candidate A appreciates that the issue is that the excess $Sn^{2+}(aq)$ would react with the potassium manganate(VII) solution and therefore make the titration impossible to interpret. Nevertheless, the answer is on the right lines, and the candidate gains 1 mark. Candidate B's answer is not the one expected, but as it is a possible problem, 1 mark is awarded.

Overall, Candidate A drops just 1 mark and scores 6 marks out of 7. Candidate B has not answered the questions with sufficient care and achieves only 2 marks.

A2 task 2: making buffer solutions

A student wants to make a buffer solution with a pH of 4.77. The student finds out that a mixture containing $0.100 \, mol \, dm^{-3}$ of ethanoic acid and $0.100 \, mol \, dm^{-3}$ of sodium ethanoate will have this pH value.

(a) Show, by calculation, that a buffer solution containing $0.100 \, mol \, dm^{-3}$ of ethanoic acid and $0.100 \, mol \, dm^{-3}$ of sodium ethanoate will have a pH of 4.77.

$(K_a = 1.7 \times 10^{-5} \, mol \, dm^{-3})$ (2 marks)

Some solid sodium carbonate and some $0.100 \, mol \, dm^{-3}$ ethanoic acid are available. The student plans first to make some $0.100 \, mol \, dm^{-3}$ sodium ethanoate solution by using the sodium carbonate and ethanoic acid.

The student suggests the following method:

- Measure $25 \, cm^3$ of $0.100 \, mol \, dm^{-3}$ ethanoic acid in a measuring cylinder and pour this into a beaker.
- Add solid sodium carbonate in small portions and stir vigorously to allow the sodium carbonate to react.
- Continue adding sodium carbonate until some solid remains unreacted in the beaker.
- Filter off the excess sodium carbonate to leave the solution, which will be $0.100 \, mol \, dm^{-3}$ sodium ethanoate.

(b) (i) Give the equation for the reaction between sodium carbonate and ethanoic acid. (1 mark)

(ii) The method planned by the student will not work. Explain why not. (2 marks)

The student then decides to make the $0.100 \, mol \, dm^{-3}$ solution of sodium ethanoate by titration. The student first makes a $0.0500 \, mol \, dm^{-3}$ solution of sodium carbonate and then, using a suitable indicator, titrates it with the $0.100 \, mol \, dm^{-3}$ ethanoic acid until the end point is reached.

(c) (i) Explain why this does not form a solution of sodium ethanoate that has a concentration of $0.100 \, mol \, dm^{-3}$. (1 mark)

(ii) What will be the concentration of the sodium ethanoate made in the student's titration? (1 mark)

A teacher tells the student that a buffer solution with a pH of 4.77 can be made from the solution of sodium carbonate and ethanoic acid using the following procedure:

- $12.5 \, cm^3$ of $0.0500 \, mol \, dm^{-3}$ solution of sodium carbonate solution should be added from a burette to $25.0 \, cm^3$ of $0.100 \, mol \, dm^{-3}$ ethanoic acid.

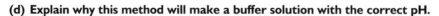

(d) Explain why this method will make a buffer solution with the correct pH.

(3 marks)

Total: 10 marks

■ ■ ■

Candidates' responses to A2 task 2

Candidate A

(a) $K_a = \dfrac{[H^+][0.1]}{[0.1]}$

$[H^+] = 1.7 \times 10^{-5}$

$pH = -\log(1.7 \times 10^{-5}) = 4.77$

Candidate B

(a) $pH = pK_a + \log\left(\dfrac{[HX]}{[X^-]}\right)$

$pH = 4.77 + \log\dfrac{[0.1]}{[0.1]} = 4.77$

Both candidates score 2 marks, although Candidate B is fortunate because the equation used to calculate pH is incorrect. The correct equation is:

$$pH = pK_a + \log\left(\dfrac{[X^-]}{[HX]}\right)$$

It is possible that 1 mark might be lost because of this, but usually if the final answer is correct the mark will be awarded. The concentrations of the ethanoic acid and the sodium ethanoate are the same so the error makes no difference to the answer.

Candidate A

(b) (i) $2CH_3COOH + Na_2CO_3 \rightarrow 2CH_3COONa + CO_2 + H_2O$

(ii) Sodium carbonate is soluble and it would be difficult to see when the right amount had been added.

Candidate B

(b) (i) $2CH_3COOH + Na_2CO_3 \rightarrow 2NaCH_3COO + CO_2 + H_2O$

(ii) The student should have looked to see when bubbling stops, so that too much sodium carbonate wasn't added.

In part (b)(i), neither candidate gives the ideal formula of sodium ethanoate, which is better written as $CH_3COO^-Na^+$. However, both obtain the mark for the equation. Neither candidate answers part b(ii) fully, though both gain 1 mark for a partly correct response. The full answer should make it clear that because sodium carbonate is soluble, once the reaction is complete it would continue to dissolve and so no solid would be visible.

Candidate A

(c) (i) When the ethanoic acid and sodium carbonate react, the solutions dilute each other.

(ii) $0.05 \, mol \, dm^{-3}$

Candidate B

(c) (i) The $mol \, dm^{-3}$ of the two solutions will have added together at the end point.

(ii) $0.150 \, mol \, dm^{-3}$

🖉 In part (c)(i), Candidate A understands what will happen to the concentration of sodium ethanoate as a result of the titration and scores the mark. The correct concentration is given in part (c)(ii), for 1 mark. Candidate B is confused and fails to score.

Candidate A

(d) During the titration, sodium carbonate reacts to form sodium ethanoate. When the titration is half complete there will be equal amounts of sodium ethanoate and ethanoic acid, which is a buffer solution. $25 \, cm^3$ of $0.100 \, mol \, dm^{-3}$ ethanoic acid gives an end point with $25 \, cm^3$ of $0.05 \, mol \, dm^{-3}$ sodium carbonate. So the volume needed is $12.5 \, cm^3$.

Candidate B

(d) When $12.5 \, cm^3$ of sodium carbonate has been added the reaction with ethanoic acid will have produced some sodium ethanoate to make the buffer solution.

🖉 Candidate A answers part (d) quite well and gets 2 of the 3 marks. For the third mark, the candidate needs a clearer statement that the buffer solution will have a pH of 4.77 when there are equal amounts, in moles, of ethanoate ion and ethanoic acid present in the same volume of solution. This will be the case when $12.5 \, cm^3$ of the sodium carbonate has been added to the ethanoic acid. Candidate A comes close to writing this but the answer is not sufficiently clear. Candidate B fails to score.

Overall, Candidate A scores 7 marks out of 10. Candidate B struggles with the more demanding parts of the assessment and only scores 4 marks.

A2 task 3: rate of reaction between magnesium and hydrochloric acid

Two students decide to investigate the rate of the reaction between magnesium and hydrochloric acid:

$$Mg(s) + 2HCl(aq) \rightarrow MgCl_2(aq) + H_2(g)$$

They carry out two experiments.

Experiment 1
- **They measure out 50 cm³ of 1.00 mol dm⁻³ hydrochloric acid using a measuring cylinder and pour it into a beaker.**
- **They weigh out 0.030 g of magnesium ribbon.**
- **They add the magnesium ribbon to the acid in the beaker and start a stopwatch.**
- **They measure the time taken for the magnesium ribbon to finish reacting. (This is shown when the ribbon is no longer visible and there are no more bubbles.)**

The students then repeat experiment 1 using 0.060 g of magnesium ribbon, instead of 0.030 g.

The students find that the time measured when 0.030 g is used is 62 s and when 0.060 g is used it is 65 s.

Experiment 2
For this experiment, the students use a similar procedure to that in experiment 1. As before, they add 0.030 g of magnesium ribbon to 50 cm³ of hydrochloric acid in a beaker and measure the time for the reaction to be completed. The experiment is carried out five times using a different concentration of hydrochloric acid each time.

The times taken for the magnesium to finish reacting are shown in the table below.

Concentration of hydrochloric acid/mol dm⁻³	Time taken for the magnesium to react/s
0.900	77.5
0.800	100
0.700	126
0.600	170
0.500	250

(a) What is the percentage error in the measurement of 50 cm³ of 1.00 mol dm⁻³ hydrochloric acid if the volume is measured with a maximum error of ±0.5 cm³? (1 mark)

(b) (i) Show by a calculation that in the reaction between 0.030 g of magnesium and 0.500 mol dm⁻³ hydrochloric acid, the hydrochloric acid is in excess. (2 marks)

(ii) Why is it essential that the acid is in excess? (2 marks)

In experiment 1, the students found that the time taken for the magnesium to react changes only very slightly when the mass of magnesium is doubled. They conclude the reaction is zero order with respect to magnesium.

(c) (i) Explain why the conclusion that the reaction is zero order with respect to the magnesium is incorrect. (1 mark)

(ii) What do the results of experiment 1 show? (1 mark)

A teacher tells the two students that the reaction is thought to be second order with respect to hydrochloric acid and asks the students to confirm whether their results support this conclusion.

(d) (i) The first student plots a graph of concentration against time and obtains a curve that does not have a constant half-life. The student concludes that the reaction is second order with respect to the hydrochloric acid.

Explain why plotting the graph and measuring the half-life is not a valid way of analysing the results of this experiment. (1 mark)

(ii) The second student decides to calculate the rate of reaction, $\frac{1}{t}$, where t is the time in seconds taken for the magnesium to react. The student then plots a graph of concentration of hydrochloric acid against $\frac{1}{t}$. The graph is shown below.

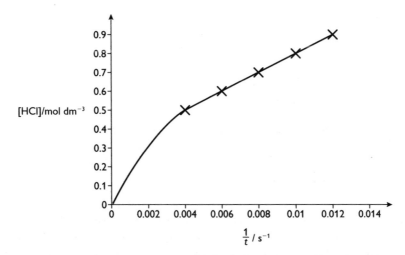

How does this graph confirm that the reaction is neither zero order nor first order? (2 marks)

(iii) To show that the reaction is second order a graph of $\frac{1}{t}$ against $[HCl]^2$ could be plotted. Describe the graph that would be obtained if the reaction is second order with respect to hydrochloric acid. (2 marks)

Total: 10 marks

■ ■ ■

Candidates' responses to A2 task 3

Candidate A

(a) Percentage error is $\left(\frac{0.5}{50}\right) \times 100 = \pm 1\%$

Candidate B

(a) Percentage error is $\left(\dfrac{0.5}{50}\right) = \pm 0.01$

 Part (a) is meant to be an easy starter. Candidate A is correct, for 1 mark. Candidate B fails to multiply by 100 to obtain the error as a percentage and therefore fails to score.

Candidate A

(b) (i) $0.030\,\text{g}$ of magnesium $= \dfrac{0.030}{24.3} = 0.00123\,\text{mol}$

$50\,\text{cm}^3$ of $1.00\,\text{mol}\,\text{dm}^{-3}$ HCl is $0.05\,\text{mol}$
$0.00123\,\text{mol}$ Mg needs $2 \times 0.00123 = 0.00246\,\text{mol}$ of HCl
So, the acid is in excess.

(ii) So that the reaction does not slow down too much during the measurement.

Candidate B

(b) (i) r.a.m. of Mg is 24.3 and $0.030\,\text{g}$ of magnesium is $0.00123\,\text{mol}$

moles of HCl are $\dfrac{50}{1000} = 0.05\,\text{mol}$

0.05 is much bigger than 0.00123, so HCl is in excess

(ii) To make sure all the magnesium has reacted.

 In part (b)(i), both candidates complete the calculations correctly and score 1 mark. Candidate A gains the second mark by relating the amounts of magnesium and hydrochloric acid to the equation. This is essential to confirm that the hydrochloric acid is in excess. Candidate B does not do this and the answer is insufficient for the mark.

Neither candidate provides the complete answer to (b)(ii). Both are awarded 1 mark although Candidate B's answer is not as good as that of Candidate A. The second mark is a difficult mark to obtain. The candidates should have mentioned that for the experiment to be valid there has to be enough hydrochloric acid for the rate to be considered to be constant throughout the course of the experiment. This is required if a comparison is to be made between the experiments.

Candidate A

(c) (i) The order of the reaction needs the concentration of the substance, but magnesium is a solid so it doesn't have a concentration.
(ii) That the surface area of the magnesium ribbon does not make much difference to the time taken for the reaction to end.

Candidate B

(c) (i) You shouldn't decide anything on the basis of only two experiments.
(ii) That in the reaction the magnesium might be zero order. You would need more experiments to be certain.

Candidate A gives an answer worth the mark for part (c)(i). Magnesium is a solid and the order requires a concentration term. Candidate A also provides a good answer to part (ii), for 1 mark. What the experiment shows is that the increase in surface area of the magnesium makes little difference when the acid is in so great an excess. Candidate B fails to score for either part although, as issues with rate experiments in general, there is some truth in the responses.

Candidate A

(d) (i)   The experiment has not been run right through its course so you can't measure half-lives. You can't prove that it is second order like this.

(ii)   A zero-order graph would be a horizontal line and a first-order graph would be a straight line through the origin. The graph plotted isn't either of those.

(iii) $\frac{1}{t}$ is the rate and rate is proportional to $[HCl]^2$, so a graph of $\frac{1}{t}$ versus $[HCl]^2$ would give a straight line going through (0,0).

Candidate B

(d) (i)   The student has only proved it is not first order and doesn't know it is second order.

(ii)   It's not zero order because the times do change. It isn't first order because the points don't all lie on the best-fit line.

(iii)   Because the 2 in HCl^2 corresponds to the 2 of second order, it would give a graph that was a straight line.

In (d)(i), Candidate B does not understand that using half-lives is incorrect. Candidate A gives the correct answer and obtains the mark. Candidate A makes a mistake in (d)(ii) by stating that the line for a zero-order reaction is horizontal. With the axes used, the line would be vertical. However, the candidate scores the mark for giving detail for a first-order reaction. Candidate B gains the mark for the answer about a zero-order reaction, but the answer for a first-order reaction is insufficient to score. Candidate A understands the reasoning behind the graph suggested in (d)(iii) and gains 2 marks. Candidate B's answer looks like a hopeful guess and the candidate fails to score.

Overall, Candidate A scores 10 marks out of 12, which is very good. Candidate B obtains only 3 marks. Although the evaluative task can be demanding, Candidate B fails to score the relatively easy marks available — perhaps through panicking when the question asked was more difficult. It should have been possible to pick up the marks for parts (a), (b) and (d) (iii), as well as at least half marks on the other part-questions. A target of 10 marks would be achievable.

Answers to questions on significant figures (p. 37)

(1) **(a)** 15.255
　　(b) 15.26
　　(c) 15.3

(2) **(a)** 0.02177
　　(b) 0.0218
　　(c) 0.022

(3) **(a)** 0.0190
　　(b) 0.019

(4) **(a)** 23.00 g
　　(b) 25.0 cm³

(5)

	Trial	1	2	3
Final burette reading/cm³	24.40	24.00	23.80	24.05
Initial burette reading/cm³	0.00	0.00	0.00	0.00
Titre/cm³	24.40	24.00	23.80	24.05
Titres used to calculate the mean/cm³		✓		✓
Mean titre value to 1 d.p./cm³	24.0			

Student Feedback Form

Philip Allan Updates would welcome your views so that we can publish exactly the right resources for you. If you complete the following form, you will be entered into a **monthly prize draw** to win a Philip Allan Updates book of your choice.

Student Unit Guide name: **OCR(A) AS/A2 Chemistry Units F323 & F326: Practical Skills in Chemistry**

Your name: Your school:

e-mail address*:

*So that we can contact you if you win the prize draw.

Why did you decide to buy this guide?

☐ Bought for me by school/college

☐ Recommended by teacher/parent/friend

☐ Found online/in a bookshop

☐ Other_____

Have you found this guide helpful for your studies / revision? _____ Yes/No

_____

How have you used this *Student Unit Guide*?

_____

_____

At what point in the course did you use this *Student Unit Guide*?

_____

_____

Is there anything else you would like from this guide?

_____

_____

What sort of online revision tools do you use?

_____

_____

Would you like to be able to buy revision guides as e-books? _____ Yes/No

Please return your completed form to:

Fax: 01869 338293

Post: Marketing Feedback, Philip Allan Updates, FREEPOST, Deddington, Oxfordshire OX14 0BR

Email: marketing@philipallan.co.uk

Internal use only:

☐ Marketing ☐ Editorial ☐ Commissioning Competition month _____